Lecture Notes in Computer Science   **12405**

More information about this series at http://www.springer.com/series/7409

Wei Song · Kisung Lee · Zhisheng Yan ·
Liang-Jie Zhang · Huan Chen (Eds.)

Internet of Things - ICIOT 2020

5th International Conference
Held as Part of the Services Conference Federation, SCF 2020
Honolulu, HI, USA, September 18–20, 2020
Proceedings

 Springer

Editors
Wei Song
North China University of Technology
Beijing, China

Zhisheng Yan
Georgia State University
Atlanta, GA, USA

Huan Chen
Kingdee International Software
Group Co. Ltd.
Hong Kong, China

Kisung Lee (iD)
Louisiana State University
Baton Rouge, LA, USA

Liang-Jie Zhang (iD)
Kingdee International Software
Group Co. Ltd.
Shenzhen, China

ISSN 0302-9743   ISSN 1611-3349   (electronic)
Lecture Notes in Computer Science
ISBN 978-3-030-59614-9   ISBN 978-3-030-59615-6   (eBook)
https://doi.org/10.1007/978-3-030-59615-6

LNCS Sublibrary: SL3 – Information Systems and Applications, incl. Internet/Web, and HCI

This Springer imprint is published by the registered company Springer Nature Switzerland AG
The registered company address is: Gewerbestrasse 11, 6330 Cham, Switzerland

Preface

With the rapid advancements of mobile Internet, cloud computing, and big data, device-centric traditional Internet of Things (IoT) is now moving into a new era which has been termed Internet of Things Services (IOTS). In this era, sensors and other types of sensing devices, wired and wireless networks, platforms and tools, data processing/visualization/analysis and integration engines, and other components of traditional IoT are interconnected through innovative services to realize the value of connected things, people, and virtual Internet spaces. The way of building new IoT applications is changing. We indeed need creative thinking, long-term visions, and innovative methodologies to respond to such a change. The ICIOT 2020 conference is organized to continue to promote research and application innovations around the world.

ICIOT 2020 is a member of the Services Conference Federation (SCF). SCF 2020 had the following 10 collocated service-oriented sister conferences: the International Conference on Web Services (ICWS 2020), the International Conference on Cloud Computing (CLOUD 2020), the International Conference on Services Computing (SCC 2020), the International Conference on Big Data (BigData 2020), the International Conference on AI & Mobile Services (AIMS 2020), the World Congress on Services (SERVICES 2020), the International Conference on Internet of Things (ICIOT 2020), the International Conference on Cognitive Computing (ICCC 2020), the International Conference on Edge Computing (EDGE 2020), and the International Conference on Blockchain (ICBC 2020). As the founding member of SCF, the First International Conference on Web Services (ICWS 2003) was held in June 2003 in Las Vegas, USA. Meanwhile, the First International Conference on Web Services - Europe 2003 (ICWS-Europe 2003) was held in Germany in October 2003. ICWS-Europe 2003 was an extended event of ICWS 2003, and held in Europe. In 2004, ICWS-Europe was changed to the European Conference on Web Services (ECOWS), which was held in Erfurt, Germany.

This volume presents the accepted papers for the International Conference on Internet of Things (ICIOT 2020), held virtually during September 18–20, 2020. Form the ICIOT 2020 conference, we accepted 12 papers for these proceedings. Each was reviewed and selected by at least three independent members of the ICIOT 2020 International Program Committee.

We are pleased to thank the authors whose submissions and participation made this conference possible. We also want to express our thanks to the Organizing Committee and Program Committee members for their dedication in helping to organize the conference and reviewing the submissions. We thank all volunteers, authors, and conference participants for their great contributions to the fast-growing worldwide services innovations community.

July 2020

Wei Song
Kisung Lee
Zhisheng Yan
Liang-Jie Zhang

Organization

General Chairs

Dimitrios Georgakopoulos   Swinburne University, Australia
Zhipeng Cai   Georgia State University, USA

Program Chairs

Wei Song   North China University of Technology, China
Kisung Lee   Louisiana State University, USA
Zhisheng Yan (Vice-chair)   Georgia State University, USA

Services Conference Federation (SCF 2020)

General Chairs

Yi Pan   Georgia State University, USA
Samee U. Khan   North Dakota State University, USA
Wu Chou   Vice President of Artificial Intelligence & Software
    at Essenlix Corporation, USA
Ali Arsanjani   Amazon Web Services (AWS), USA

Program Chair

Liang-Jie Zhang   Kingdee International Software Group Co., Ltd, China

Industry Track Chair

Siva Kantamneni   Principal/Partner at Deloitte Consulting, USA

CFO

Min Luo   Georgia Tech, USA

Industry Exhibit and International Affairs Chair

Zhixiong Chen   Mercy College, USA

Operations Committee

Jing Zeng   Yundee Intelligence Co., Ltd, China
Yishuang Ning   Tsinghua University, China
Sheng He   Tsinghua University, China
Yang Liu   Tsinghua University, China

Steering Committee

Calton Pu (Co-chair)   Georgia Tech, USA
Liang-Jie Zhang (Co-chair)   Kingdee International Software Group Co., Ltd, China

ICIOT 2020 Program Committee

Georgios Bouloukakis   University of California, Irvine, USA
Luca Cagliero   Politecnico di Torino, Italy
Tao Chen   University of Birmingham, UK
Nagarajan Kandasamy   Drexel University, USA
Rui André Oliveira   University of Lisbon, Portugal
Françoise Sailhan   CNAM, France

Conference Sponsor – Services Society

Services Society (S2) is a nonprofit professional organization that has been created to promote worldwide research and technical collaboration in services innovation among academia and industrial professionals. Its members are volunteers from industry and academia with common interests. S2 is registered in the USA as a "501(c) organization," which means that it is an American tax-exempt nonprofit organization. S2 collaborates with other professional organizations to sponsor or co-sponsor conferences and to promote an effective services curriculum in colleges and universities. The S2 initiates and promotes a "Services University" program worldwide to bridge the gap between industrial needs and university instruction.

The services sector accounted for 79.5% of USA's GDP in 2016. The world's most service-oriented economy, with services sectors accounting for more than 90% of GDP. S2 has formed 10 Special Interest Groups (SIGs) to support technology and domain specific professional activities:

- Special Interest Group on Web Services (SIG-WS)
- Special Interest Group on Services Computing (SIG-SC)
- Special Interest Group on Services Industry (SIG-SI)
- Special Interest Group on Big Data (SIG-BD)
- Special Interest Group on Cloud Computing (SIG-CLOUD)
- Special Interest Group on Artificial Intelligence (SIG-AI)
- Special Interest Group on Edge Computing (SIG-EC)
- Special Interest Group on Cognitive Computing (SIG-CC)
- Special Interest Group on Blockchain (SIG-BC)
- Special Interest Group on Internet of Things (SIG-IOT)

About the Services Conference Federation (SCF)

As the founding member of the Services Conference Federation (SCF), the First International Conference on Web Services (ICWS 2003) was held in June 2003 in Las Vegas, USA. Meanwhile, the First International Conference on Web Services - Europe 2003 (ICWS-Europe 2003) was held in Germany in October 2003. ICWS-Europe 2003 was an extended event of ICWS 2003, and held in Europe. In 2004, ICWS-Europe was changed to the European Conference on Web Services (ECOWS), which was held in Erfurt, Germany. SCF 2019 was held successfully in San Diego, USA. To celebrate its 18th birthday, SCF 2020 was held virtually during September 18–20, 2020.

In the past 17 years, the ICWS community has been expanded from Web engineering innovations to scientific research for the whole services industry. The service delivery platforms have been expanded to mobile platforms, Internet of Things (IoT), cloud computing, and edge computing. The services ecosystem is gradually enabled, value added, and intelligence embedded through enabling technologies such as big data, artificial intelligence (AI), and cognitive computing. In the coming years, all the transactions with multiple parties involved will be transformed to blockchain.

Based on the technology trends and best practices in the field, SCF will continue serving as the conference umbrella's code name for all service-related conferences. SCF 2020 defines the future of New ABCDE (AI, Blockchain, Cloud, big Data, Everything is connected), which enable IOT and enter the 5G for Services Era. SCF 2020's 10 collocated theme topic conferences all center around "services," while each focusing on exploring different themes (web-based services, cloud-based services, big data-based services, services innovation lifecycle, AI-driven ubiquitous services, blockchain driven trust service-ecosystems, industry-specific services and applications, and emerging service-oriented technologies). SCF includes 10 service-oriented conferences: ICWS, CLOUD, SCC, BigData Congress, AIMS, SERVICES, ICIOT, EDGE, ICCC, and ICBC. The SCF 2020 members are listed as follows:

[1] The International Conference on Web Services (ICWS 2020, http://icws.org/) is the flagship theme-topic conference for Web-based services, featuring Web services modeling, development, publishing, discovery, composition, testing, adaptation, delivery, as well as the latest API standards.

[2] The International Conference on Cloud Computing (CLOUD 2020, http://thecloudcomputing.org/) is the flagship theme-topic conference for modeling, developing, publishing, monitoring, managing, delivering XaaS (Everything as a Service) in the context of various types of cloud environments.

[3] The International Conference on Big Data (BigData 2020, http://bigdatacongress.org/) is the emerging theme-topic conference for the scientific and engineering innovations of big data.

[4] The International Conference on Services Computing (SCC 2020, http://thescc.org/) is the flagship theme-topic conference for services innovation lifecycle that includes enterprise modeling, business consulting, solution creation, services orchestration,

services optimization, services management, services marketing, and business process integration and management.

[5] The International Conference on AI & Mobile Services (AIMS 2020, http://ai1000. org/) is the emerging theme-topic conference for the science and technology of AI, and the development, publication, discovery, orchestration, invocation, testing, delivery, and certification of AI-enabled services and mobile applications.

[6] The World Congress on Services (SERVICES 2020, http://servicescongress.org/) focuses on emerging service-oriented technologies and the industry-specific services and solutions.

[7] The International Conference on Cognitive Computing (ICCC 2020, http:// thecognitivecomputing.org/) focuses on the Sensing Intelligence (SI) as a Service (SIaaS) which makes systems listen, speak, see, smell, taste, understand, interact, and walk in the context of scientific research and engineering solutions.

[8] The International Conference on Internet of Things (ICIOT 2020, http://iciot.org/) focuses on the creation of Internet of Things (IoT) technologies and development of IoT services.

[9] The International Conference on Edge Computing (EDGE 2020, http:// theedgecomputing.org/) focuses on the state of the art and practice of edge computing including but not limited to localized resource sharing, connections with the cloud, and 5G devices and applications.

[10] The International Conference on Blockchain (ICBC 2020, http://blockchain1000. org/) concentrates on blockchain-based services and enabling technologies.

Some highlights of SCF 2020 are shown below:

- **Bigger Platform:** The 10 collocated conferences (SCF 2020) are sponsored by the Services Society (S2) which is the world-leading nonprofit organization (501 c(3)) dedicated to serving more than 30,000 worldwide services computing researchers and practitioners. Bigger platform means bigger opportunities to all volunteers, authors and participants. Meanwhile, Springer sponsors the Best Paper Awards and other professional activities. All the 10 conference proceedings of SCF 2020 have been published by Springer and indexed in ISI Conference Proceedings Citation Index (included in Web of Science), Engineering Index EI (Compendex and Inspec databases), DBLP, Google Scholar, IO-Port, MathSciNet, Scopus, and ZBlMath.
- **Brighter Future:** While celebrating the 2020 version of ICWS, SCF 2020 highlights the Third International Conference on Blockchain (ICBC 2020) to build the fundamental infrastructure for enabling secure and trusted service ecosystems. It will also lead our community members to create their own brighter future.
- **Better Model:** SCF 2020 continues to leverage the invented Conference Blockchain Model (CBM) to innovate the organizing practices for all the 10 theme conferences.

Contents

Image Privacy Protection by Particle Swarm Optimization Based Pivot
Pixel Modification.. 1
    Jishen Yang, Yan Huang, Junjie Pang, Zhenzhen Xie, and Wei Li

On Burial Depth of Underground Antenna in Soil Horizons
for Decision Agriculture... 17
    Abdul Salam and Usman Raza

Deriving Interpretable Rules for IoT Discovery Through Attention 32
    Franck Le and Mudhakar Srivatsa

Combining Individual and Joint Networking Behavior for Intelligent
IoT Analytics 45
    Jeya Vikranth Jeyakumar, Ludmila Cherkasova, Saina Lajevardi,
    Moray Allan, Yue Zhao, John Fry, and Mani Srivastava

RelIoT: Reliability Simulator for IoT Networks...................... 63
    Kazim Ergun, Xiaofan Yu, Nitish Nagesh, Ludmila Cherkasova,
    Pietro Mercati, Raid Ayoub, and Tajana Rosing

NACK-Based Reliable Multicast Communication for Internet of Things
Firmware Update 82
    Jiye Park, Dongha Lee, Markus Jung, and Erwin P. Rathgeb

A WiVi Based IoT Framework for Detection of Human Trafficking Victims
Kept in Hideouts... 96
    Sidharth Samanta, Sunil Samanta Singhar, A. H. Gandomi,
    Somula Ramasubbareddy, and S. Sankar

Spatio-Temporal Coverage Enhancement in Drive-By Sensing Through
Utility-Aware Mobile Agent Selection 108
    Navid Hashemi Tonekaboni, Lakshmish Ramaswamy, Deepak Mishra,
    Omid Setayeshfar, and Sorush Omidvar

Development of an Electronic System for the Analysis and Integration
of Data on Water Care.. 125
    Febe Hernandez Aleman and Martha S. Lopez-de la Fuente

BWCNN: Blink to Word, a Real-Time Convolutional Neural
Network Approach 133
    Albara Ah Ramli, Rex Liu, Rahul Krishnamoorthy, I. B. Vishal,
    Xiaoxiao Wang, Ilias Tagkopoulos, and Xin Liu

Risk Assessment of Vehicle Sensor Data as a Vending Object or Service   141
    Frank Bodendorf and Jörg Franke

Review for Message-Oriented Middleware   152
    Yang Liu, Liang-Jie Zhang, and Chunxiao Xing

Author Index   161

Image Privacy Protection by Particle Swarm Optimization Based Pivot Pixel Modification

Jishen Yang[1], Yan Huang[2(✉)], Junjie Pang[3], Zhenzhen Xie[4], and Wei Li[1]

[1] Department of Computer Science, Georgia State University, Atlanta 30303, USA
[2] Department of Software Engineering and Game Development, Kennesaw State University, Atlanta 30144, USA
yhuang24@kennesaw.edu
[3] College of Computer Science and Technology, Qingdao University, Qingdao 266000, China
[4] College of Computer Science and Technology, Jilin University, Changchun 130012, China

Abstract. The image classification models based on neural networks recently have outperformed most of the traditional models, and rapidly been developed and implemented by industry because of the capability of qualifying various computer vision tasks. Hence, the exposure of users' image data to unauthorized powerful models causes more information leak in a shorter time. Through experiments, we find that for each input image, the change of the image's prediction scores by each pixels' RGB value change is different. Also, the pattern of the sensitivity on each pixel is highly related to the category and composition of the input image. By utilizing this feature, we present Pivot Pixel Noise Generator by Particle Swarm Optimization to generate noise points on original images to lower the target model's accuracy of correctly predicting the target image's label, so to protect the information contained in the target image from the image classification models. The model performs in a semi-black-box manner and balances the number of queries to the target and total number of modified points. We also propose an initialization strategy for the model, PSO Knowledge Transfer, which initializes the model's parameters with experience learned from previous runs to further reduce the number of query times and noise points. The model is evaluated using the image classification benchmark model $ResNet50$ and shows an advantage compared to the baseline algorithm.

1 Introduction

Deep learning models have been significantly developed in recent years. The robust performance of deep learning models inspired various applications in various industrial fields. Besides the escalation of efficiency, artificial intelligence models require a large amount of data to be trained and how to reserve privacy in deep learning becomes a severe issue. The main contributors to data booming,

© Springer Nature Switzerland AG 2020
W. Song et al. (Eds.): ICIOT 2020, LNCS 12405, pp. 1–16, 2020.
https://doi.org/10.1007/978-3-030-59615-6_1

including social networks and Internet of Things (IoT), usually contain users' sensitive private information. While, existing approaches that aim to protect privacy of big data are either to apply privacy protection on the original user data [1,2], on the way of data being transferred [3–5], or at the stage of data being used by deep learning networks [6,7].

In the field of computer vision, deep learning neural networks developed dramatically since 2012. Alexnet [8] was introduced to the ImageNet Large Scale Visual Recognition Challenge, which is one of the most important benchmarks in computer vision. Alexnet utilize the idea of convolutional neural network and achieves the highest performance of accuracy in the task of image classification and motivates a number of related researches. Indeed, the image classification models facilitate the work of visual data analysis, but the potential of undesired use of image classification models increases the risk of information leaking. Once the private pictures are exposed to those image classification models, the automatic algorithms can add labels to the visual data in a short time period with high accuracy. Then the malicious party could extract more sensitive information from the visual data analysis. With the improvement of IoT sensors and publicly accessible cameras, visual data becomes one of the new leading sources of privacy loss [9]. For example, if personal photos are leaked to a malicious party, the party could further extract the personal information contained in the photos, such as location information from a photo with a landmark building, financial information from a photo of a valued car, demographic information like gender and age from a selfie, and even personality and consumer preferences with more sophisticated analysis. Or, if pictures of business files are leaked, the loss could be more critical. Since all the extraction of features from the image are automatic with high speed and information clips from each individual image could be integrated in a timely manner, the data leak could cause significant further cost. In this way, to better protect the confidential content of visual data even compromised to unauthorized access, an effective information encryption model for artificial intelligence models is necessary. Also, with the development of research on machine learning, some popular cloud service providers, such as Amazon [10], Google [11], and Microsoft [12], offer a machine learning service solution: Machine Learning as a Service (MLaaS). The MLaaS platforms set up machine learning environments and computation resources. And the training dataset can include private information. Song et al. [13] found that even the malicious party only has access to the parameters of the machine learning model, they still can infer information of the training set.

To preserve privacy and information security, the raw image data either stored locally or on the cloud needs protection from convolutional neural networks. Although the convolutional neural networks achieved much progress in computer vision, the model structure is vulnerable to minor changes in value at critical locations on the original input image data, and the output of the networks can be strongly influenced by even minor perturbations. In 2019, Su et al. [14] proposed a novel adversarial example generating model to conduct attacks on convolutional neural networks by changing only several or even just one pixel's RGB value in an input image. The attack model is named One Pixel Attack and it demonstrates impressive performance by adding noise on one pixel level to original input image data to interference the prediction of target classification neural network. The attack is in a semi-black-box manner. The target model's structure and parameters are not accessible, and the attack model has unlimited opportunities of attempts of sending modified image to and collecting output from target model. However, the One Pixel Attack is a more heuristic idea than practical solution, since it only relies on a relatively low-resolution image input setting, which is 32 by 32 pixels, the model illustrates best performance. To conduct attack on high-resolution image, the attack needs a longer length of perturbation that is an acceptable larger number of points where the noise is added on. Then the proposed optimization method, Deferential Evolution, shows lack of efficiency and automaticity because the new attempts of where to add the noise are pure randomly selected, and the number of attacked points is determined by manually set hyper-parameters.

Thus, with the concept of adversarial example attack, we propose a novel noise generation model to prevent malicious machine learning powered computer vision models gaining too much information from users' image data. Our data protection model Pivot Pixel Noise Generator (PPNG), with Particle Swarm Optimization (PSO) [15] provides protection to image data with noises on a small number of pixels on high-resolution image from classification neural networks. The proposed model utilizes PSO Knowledge Transfer, an initialization strategy of PSO parameters with the experiences from previous target images. Also, the model can automatically self-adapt the number of points to be modified, ensuring the efficiency of protection. In other words, the model can perform pixel-level attack with vastly reduced queries to the target image classification model, meanwhile maintaining a tolerable trade-off of a modestly increased number of affected pixels.

The main contributions of this work are highlighted as follows:

- We achieve privacy protection on image data from image classification neural networks by reducing the prediction accuracy with a proposed complete half-black-box privacy protection model, Pivot Pixel Noise Generator, which balances the trade-off of running time and number of pixels modified.
- We define the sensitivity map of images to image classification neural networks. Each input image has a sensitivity map which potentiates the protection strategy of locating and modifying critical points to alter classification results without knowledge of target model's architecture and parameters. Through the features of sensitivity maps, we propose to utilize PSO to optimize our model.
- We introduce a novel strategy of PSO parameters initialization, Knowledge Transfer. On each time of noise-generation on target images, the model keeps the experience of PSO parameters for future uses, which significantly reduces the running time and the number of queries to the target model, and hence facilitates the processing of protection on a large amount of image data.

2   Related Works

The research on adversarial example attack in image classification neural network models has been studied for years. In 2014, Szegedy et al. [16] discovered that many state-of-the-art neural networks are sensitive to *adversarial examples*. The term *adversarial examples* are the examples images based on original example images with only slightly perturbations which are hardly perceivable by human observers but cause the neural network models to misclassify the example to wrong categories. In 2015, Nguyen et al. [17] introduced a novel approach to generate images to fool the deep learning classifiers. The evolutionary algorithm generated images are not recognizable by human, and the target neural networks misclassify the images to wrong categories with very high confidence. This work reveals that deep learning models are very vulnerable to intentionally built adversarial examples. However, the method is not to protect existing image data. Papernot et al. [18] in their 2016 work discussed the relationship of attack difficulty and information known about the target model. The paper introduces a concept of *Saliency Map*, which summarizes the performance change of the target model to the location where the noise is added but computed with information of the target model's architecture and weight parameters.

Papernot et al. [19] in 2017 introduced a new idea of performing an adversarial example attack in a black-box manner. They managed to build a substitute model in order to overpass the need to acquire the target model's parameter and architecture. The work is tested on MNIST [20] and GTSRD [21] dataset, which relatively lower resolution level. When the input images have a larger size, and the target model has more complicated architecture, the cost of training the attack model increases rapidly. Su et al. [14] developed their adversarial example attack model in a more extreme setting that the adversarial example is only one pixel different from the normal example in value. However, the model only uses Differential Evolution as the optimizer, and hence the attack model does not sufficiently utilize the information of previous attack attempts. Also, the model illustrates a lower attack successful rate on *ImageNet* dataset than *Cifar*10 [22], which indicates that for examples of more pixels, noise with a length of only one pixel is not enough.

Zhang et al. [23] developed an adversarial example attack model based on PSO, and the setting of the particles is the same size of the target image. The optimization goal is to alter the predicted label with a limitation of $L2$ distance between the modified target image and the original image. Similarly, Mosli et al. [24] show their attack model based on PSO, with each particle initialized as a target image with randomized noise added on. Further, they proposed a procedure to reduce excess perturbations when the attack is already successful in avoiding the adversarial examples that differ too much from the original examples. However, these applications of PSO ignore previous experiences.

3   System Model

The generation of the protective noises on original images can be formalized as an optimization problem. The optimization goal is to minimize the output classification score of the correct label while maintaining a minimum number of pixels affected. The PPNG model is consisted by 3 components, including Point Value (find the sensitivity of a certain pixel), PSO Update (find the pivot pixel on current stage of protection), and Pivot Pixel Noise Generator (add noise on the pivot pixel that PSO has found).

3.1   Model Architecture

In our settings, the pixel noise generator is the function to be optimized. The pixel noise generator works in the way that the input is the location of the pixel to add noise on, and the output is the new prediction score of the modified image by the target image classification model. To maximize the performance of the privacy protection model, we use PSO to optimize the function, i.e., to find the best position of the pixel to add noise on, which ensures the target model returns global minimum output classification score for the correct class with only one pixel's RGB value been changed. For this point, each particle has two dimensions, which represent the location of the pixel to be modified in the original input image to find the most influential pixel on the current image, which can be defined as the pivot pixel.

PSO is an optimizer that fits well for the setting. It is free from requiring the full computation details from the target model and is powerful for finding the global and local minimum of a function. PSO works with a population of candidate solutions, in another term, the swarm of particles. Each particle is a candidate solution in the searching space, which has the same dimension of an input to the function to be optimized. The particles move with speed and direction based on each particle's own known best position and global-best-known position of the entire swarm. After each iteration, the best-known positions of each particle and the entire swarm are updated by queries to the target function, and the current position of the particle is updated by calculated movement speed and direction.

However, the raw images taken into the target model have a size of 224 by 224 pixels, which is 49 times more than images of 32 by 32 pixels in $Cifar10$ dataset. Only changing one single pixel's value is not enough to affect the classification scores by enough amount to alter the top 1 predicted class. To further improve the protection performance, more pixels need to be modified, and to continue implementing the modification on more pivot pixels, we define a new term *round* which means that for each round the pixel noise generator is going to modify one pivot pixel whose position found by PSO after T iterations. By the end of each round, the pixel level of noise is added to the global-best point, which is supposed to be the pivot point found by PSO. Then, the next round of optimization will be conducted on the newly modified image. The most notable innovation of the optimization is that between each round, all of the temporary parameters learned from previous round can be used for the new round of searching the pivot point on the new state of the image. Since the gradient maps of the target image with p points modified and that with $p + 1$ points modified have similar patterns, the experiences of PSO learned from previous rounds are fit for the new round. Under this framework, the number of queries to the target model is significantly dropped, and the privacy protection model can perform faster with the noise adding position still near pivot point in a desirable trade-off. If the protection is successful in R rounds of optimization, it means that the privacy protection model is able to alter the classification result to incorrect labels by modifying R pixels on the original image. Furthermore, the Knowledge Transfer of PSO works not only on the consecutive rounds of optimization on the same image but also on a new image in the same category.

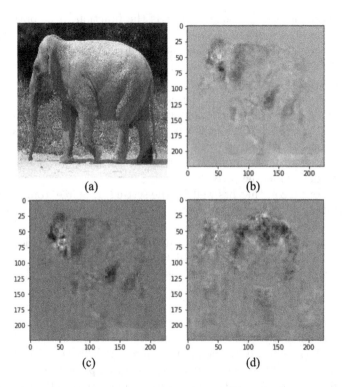

Fig. 1. (a) is the original input image *elephant*1, and (b) is the sensitivity map of the original image to target model *ResNet*50 [25]. (c) is the sensitivity map of the image modified with one pixel from the original image. (d) is the sensitivity map of another image *elephant*2 in the same class.

In the PPNG model, the feasibility of the utilization of optimizer by PSO and Knowledge Transfer of PSO is based on three assumptions. The first assumption is that the sensitivity map of a target image has gradients and suitable for PSO to find global maximum point. The sensitivity of a pixel is the decrease of the output confidence score of the corresponding correct label of image classification models with the image if the pixel is modified with noise. The sensitivity map of an image is a map with the same size as the original image, and the value of each pixel is the pixel's sensitivity. The second assumption is that the sensitivity maps of two images different from one pixel, or in other words, two images from two consecutive rounds are similar in pattern. The third assumption is that the sensitivity maps of two different images from the same category or similar composition are similar in pattern.

To test and verify the feasibility of our model, we conducted the experiments to validate our assumptions. The target model is $ResNet50$ trained with $ImageNet$ dataset [26], and the input images are exclusive from the training dataset. The experiment is to add noise on every single pixel independently from the rest of all pixels and send each of the modified new images to the target model to get predict results. Then, we assign each pixel's sensitivity as the amount of decrease in the classification score of the correct class label from collected results. Finally, we build the sensitivity map of each test image for different target models.

In (b) of Fig. 1, the sensitivity map clearly shows that PSO is suitable for the function because the function values change smoothly without very sharp outliers. Further, we obtain and compare the sensitivity maps of the same image with only one pixel in difference. As the results show in (c), the similarity in pattern ensures that our setting of inheriting PSO parameters from previous rounds is valid. Last, in (d) of Fig. 1, the sensitivity maps of two images with analogous structure help the assumption 3 stand.

3.2 Problem Description

The optimization of generating adversarial example of images is to maximize the decrease in prediction accuracy of target image classification model which under constraints. We formalize the input image as a vector x with length or dimensions as n, and each of the elements is one pixel. The target model noted as f, which takes input of n-dimensional vector $x = (x_1, x_2, \ldots, x_n)$, and output a classification label and score, L and $f_L(x)$. $f_L(x)$ also represents the probability of x falling into the class L. We define the noise vector, which is to be added on the original input vector x as vector $e(x)$. In our case, we aim to maximize the decrease of $f_L(x')$ and $x' = x + e(x)$, while minimize the length of $e(x)$.

$$\text{maximize} f_{notL}(x + e(x)) \tag{1}$$

$$\text{minimize} ||e(x)|| \tag{2}$$

3.3 Model Components

The formal algorithm shows below. For the Algorithm 1: Point Value, the input P is a list with a length of 2, representing the coordinates on axles. f is the image classification model that the protection model is aiming to duel against. The input image x with correct label L is the target image. The return value of model f with input x is presented as $f(x)$, which is a list of scores for different predicted categories, and $f_L(x)$ is the confidence level of x fall in the correct category L. In other terms, this algorithm is to add noise on the point P and returns the decrease of $f_L(x)$.

Algorithm 1: Point Value.

Input: P, x, L, and f.

$x_{modified} \leftarrow x +$ Noise at Point P;

$PointValue \leftarrow f_L(x) - f_L(x_{modified})$;

Return $PointValue$

For the Algorithm 2: PSO Update Particle Positions, this is a PSO optimizer application in setting of each particle with dimension of 2, and the value of a particle position is calculated by Algorithm 1. We set the PSO with a total of n particle in the swarm and initialization of start positions G and velocities V of particles uses knowledge transfer from selected images of a similar pattern. *pbest* represents each particles' history position of best value, and *gbest* is the position of best value of the whole swarm. ks are the multiplier constants for calculating accelerations of velocity for positions of particles in the next iteration, where k_1 is for the part of acceleration based on each particle's own experience, and k_2 is for that based on the experience of the society of swarm.

Algorithm 2: PSO Update Particle Positions.

Input: *pbest*, *gbest*, V, G, n, k_1, and k_2.

for $i \leftarrow 0$ **to** n **do**

$\quad V_i \leftarrow$

$\quad\quad V_i + k_1 * uniform(0, 1) * (pbest_i - G_i) + k_2 * uniform(0, 1) * (gbest - G_i)$

$\quad G_i \leftarrow G_i + V_i$

$\quad CurrentPointValue \leftarrow PointValue(G_i, x)$

$\quad ParticleBestValue \leftarrow PointValue(pbest_i, x)$

$\quad$ **if** $CurrentPointValue > ParticleBestValue$ **then**

$\quad\quad |\quad pbest \leftarrow G_i$;

$\quad$ **end**

end

Algorithm 3: Pivot Pixel Noise Generator by PSO is our integrated model. For each round, PSO repeats to optimize the function and find the pivot point of best protection performance. If the output score of correct class from target model with modified image is still the highest among scores of all classes, which means the protection is not yet successful, and for continuous T iterations, the decrease of output score of correct class is minor than a threshold h, the algorithm moves to a new round. The hyper-parameter h and T balance the accuracy of searching the actual pivot point and the running time of the algorithm. If h is set to 0, and T set to infinity, the algorithm is always to find the actual pivot point, but the running time compromises significantly and hence inefficient. The point on the target image at position *gbest* is then modified at end of each rounds, and the target image is replaced by the newly modified image for the next round.

Between two consecutive rounds, the parameters of PSO are inherited, including G, V, *pbest* and *gbest*. If the protection succeeded, R is the count of pixels where all the noises are added to.

Algorithm 3: Pivot Pixel Noise Generator by PSO.

Input: x, L, n, T, h, k_1, k_2, f, *pbest*, *gbest*, G, and V.
initialization:
$R \leftarrow 0$;
$ModifiedImage \leftarrow x$;
while $Max(f(ModifiedImage)) = f_L(ModifiedImage)$ **do**
    PSO Update Particle Positions
    for $i \leftarrow 0$ **to** n **do**
        $ParticleBestValue \leftarrow PointValue(pbest_i, ModifiedImage)$
        $GlobalBestValue \leftarrow PointValue(gbest, ModifiedImage)$
        if $ParticleBestValue > GlobalBestValue$ **then**
            $gbest \leftarrow pbest_i$;
        end
    end
    if $GlobalBestValue[-T] - GlobalBestValue[-1] < h$ **then**
        $ModifiedImage \leftarrow ModifiedImage +$ Noise at Point *gbest*;
        $R \leftarrow R + 1$
    end
end
Return$ModifiedImage$, R

Algorithm 4: Exhaustive Search.

Input: x with height H and weith W, L, and f
initialization:
$TempScore \leftarrow 0$;
$BestScore \leftarrow 1$;
$BestPoint \leftarrow (0,0)$;
$ModifiedImage \leftarrow x$;
while $Max(f(ModifiedImage)) = f_L(ModifiedImage)$ **do**
    for $h \leftarrow 0$ **to** H **do**
        for $w \leftarrow 0$ **to** W **do**
            $TempImg \leftarrow ModifiedImage$;
            Add Noise on $TempImg$ at (h, w);
            $TempScore \leftarrow f_L(TempImg)$;
            if $BestScore > TempScore$ **then**
                $BestScore \leftarrow TempScore$;
                $BestPoint \leftarrow (h, m)$;
            end
        end
    end
    Add Noise on $ModifiedImage$ at $BestPoint$;
end
Return $ModifiedImage$

The baseline algorithm towards this problem is the exhaustive search. We design the exhaustive search algorithm for a minimum number of modified points to alter the target model's classification top 1 category as in Algorithm 4. The basic idea of the exhaustive search algorithm is to run a noise adding test on every point of the image to the target model, and always modify the most influential point until the predicted category is changed.

4   Experiments

To test our the privacy protection model, we set up simulation in the following settings. We use *ResNet*50 as the target model, the image classification model to be protect from, with parameters trained on *ImageNet* dataset taking input image size of 224 by 224 pixels. In the first simulation, we select the same test picture *elephant*1 showing an elephant. The target model's original top 3 classification results of the image are 'Indian elephant', 'tusker', and 'African elephant' with confidence level of 0.954, 0.024 and 0.017 accordingly.

The test results of the baseline algorithm of exhaustive search show that by always selecting and modifying the most influential pixel (in other words, the point which has the highest value on the sensitivity map) on each state of noise adding process, at least 16 points are necessary to be altered in order to change the predicted category. As shown in (a) of Fig. 2, the new results of the modified image are top 1 category of 'Arabian camel' with a probability of 0.303.

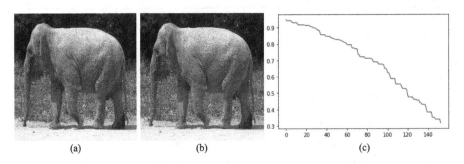

(a)   (b)   (c)

Fig. 2. (a) is the modified image by exhaustive search algorithm. (b) is the modified image by PPNG model. (c) is the curve of output confidence score of the category 'Indian elephant' by iterations.

Although the protection using the exhaustive search algorithm is successful, the protection procedure takes an enormous amount of time and computation power, and most importantly great number of queries to the target model. To decide one location of a pivot pixel, the algorithm needs to access the target model by $224 * 224$ times. The target model is called $16 * 224 * 224$ times, and thus it is not ideally efficient.

Then, we test PPNG model on the same image to the same target model. The hyper-parameters are set as acceleration constants $k1 = k2 = 0.1$, particle number $n = 50$, score decrease threshold $h = 0.01$ and iteration number $T = 5$. The initialization of *pbest*, *gbest* and G as random numbers with upper bound of 223 and lower bound of 0, and V as 0s. Then, our model can successfully conduct the protection by modifying 22 points with the total number of iteration of PSO as 152. So, the new protection model dramatically reduces the number of accessing to the target model to $152 * 50$ times. In this manner, the protection model sacrifices the number of modified pixels and save the running time and target model queries with ensuring protection success. The protection results of the model to the modified image in (b) of Fig. 2 are also top 1 category of 'Arabian camel' with a probability of 0.317. In this way, the image classification model is no more able to correctly extract information from the target image, and thereby the privacy is preserved. Additionally, in (c) of Fig. 2 the curve of the performance of the protection indicates the strategy of setting the threshold h and iteration number T effectively saves the protection model from over-fitting and so reduces the running time. The curve is the output confidence score of the correct category by each time of updating with PSO. The x axle is the number of iterations, and the y axle is the confidence score of the ground truth category. With h set to 0.01 and T set to 5, if for five consecutive iterations, all of the particles in the swarm have not found a better position, the pixel according to the *gbest* is modified with noise. The horizon part of the ladder-shaped curve indicates that the numbers of iterations of *gbest* not been updated, and the fast drop indicates that a new noise point is made. Specifically in iteration 52 to 60, the PSO keeps updating the *gbest*, and from iteration 60 to 64 the *gbest* remains the same value. So, after the iteration of 64, with a new noise point added to the target image, the output score of the correct label decreased quickly, which means that PSO is able to find the next pivot point very fast. During iteration 100 to 120, again the PSO shows the capability of locating the pivot point on updated target image.

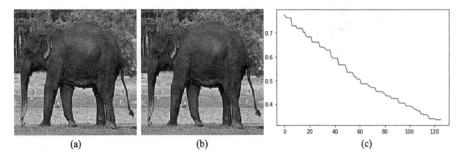

(a)   (b)   (c)

Fig. 3. (a) is the adversarial example generated by exhaustive search with 11 points modified, (b) is by PPNG model with 19 points modified, and (c) is the curve of score of label 'Indian elephant', all based on image '*elephant2*'.

This simulation illustrates the advantage of PPNG model. The PSO can find or approach the global-optimal point on the sensitivity map without running a test on every pixel. And the design of setting a threshold of optimization of each round helps the PSO to prevent excessive updates of particle positions and queries to the target model because the difference in the performance of the current *gbest* and optimal point of actual sensitivity map is acceptable considering the cost of computation and number of queries by continuing updating. The efficiency of lower the accuracy of prediction of target model enables the model to provide protection to images in a short time.

The simulations on other images in the same category show similar results. Image *elephant*2 has original top 3 classification results of 'Indian elephant', 'tusker', and 'African elephant' with a confidence level of 0.803, 0.147, and 0.047. By exhaustive search algorithm, the image needs 11 points to be modified to change the output category. The number of accessing the target model is $11 * 224 * 224$. By using the PPNG model with random initialization and the same hyper-parameters as simulation of image *elephant*1, the number of noise-added points increases to 19 but only after 125 update iterations, which makes $125 * 50$ queries. The modified results are in Fig. 3

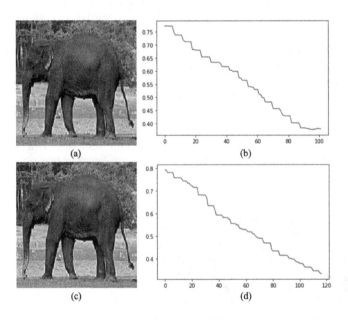

(a)   (b)

(c)   (d)

Fig. 4. (a) and (c) are the modified image by PPNG model initialized knowledge transferred from the protection procedure of image *elephant*1 on the same target model *ResNet*50 and knowledge transferred from the protection procedure of image *elephant*2 on *VGG*16. (b) and (d) are the corresponding score curves.

After the first protection conducted for images in one category, the model saves the previous experiences for knowledge transfer of PSO for future protection initialization use of similar images. We test the PPNG model on the same image with knowledge from the previous test image *elephant*1 on the same target model. The protection goes successful by changing the predicted category to 'tusker' with a confidence level of 0.380, with not only the number of points altered dropping to 16 but also iteration number reducing to 101. Moreover, initialization with knowledge from the same image of *elephant*2 on different target image classification model $VGG16$ [27] also helps. With knowledge transfer, the protection is done with top 1 predicted category of 'tusker' with a confidence level of 0.334. The number of altered points keeps 19, and the number of total iterations of PSO reduces to 116. The results in Fig. 4 illustrate the advantage of knowledge transfer. At early stage, the PSO locates the global-best position of particles in a faster manner on the new image, and the curve of confidence level shows more of ladder-shaped drops in the first iterations, which indicates that the PSO finds the global-best position more swiftly and clearly.

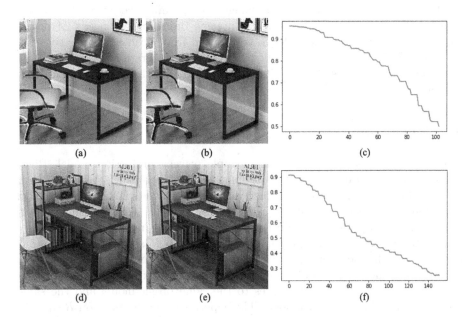

Fig. 5. (a) and (d) are the modified results by exhaustive search, (b) and (e) are by PPNG model, (c) and (f) are the curve of score of label 'desk', of image *'desk*1' and *'desk*2'.

Similarly, we test more images with different categories. The original image *'desk*1' has the classification result of 'desk' with the confidence level of 0.962, and 'dining table' with 0.032 and 'desktop computer' with 0.001. By exhaustive search algorithm, the target model gives the results of 'dining table' with a confidence level of 0.554, which becomes the top 1 classification category. The target image is added by noise at 11 points, and the exhaustive search algorithm queries the target model by $11 * 224 * 224$ times. By using the PPNG model with the same hyper-parameter setting, the queries to target model reduce to $103 * 50$ times and with total 16 points modified to change the top 1 predicted category to 'dining table' with a score of 0.500. For test image '$desk2$', the original score is 0.938, and the baseline algorithm needs 13 points and $13 * 224 * 224$ calls to the target model, while PPNG model needs 24 points with $151 * 50$ queries. The results are shown in Fig. 5. The reason why the PPNG model needs more points is that for these two images, the sensitivity map is not very evident with gradients. The surroundings of the table also show the importance of affecting the target model output results. As shown in (d) of Fig. 5, the baseline algorithm chooses points near the edge of the image, where the pixel shows the largest influence on the target model performance. Also, the difference of sensitivity of pixels is not as significant as those of images in other categories. But still, the PPNG model managed to provide protection with substantially fewer queries to the target model.

Additionally, the protection on image '$desk2$' initialized with knowledge transferred from protection of '$desk1$' on same target model, needs 24 points and 142 iterations. With knowledge transferred from itself on alternative target model $VGG16$, PPNG needs 17 points with 106 iterations of update. The improvement illustrates the benefits of initialization with knowledge than randomized numbers.

5   Conclusion

This paper proposes an adversarial example privacy protection model, Pivot Pixel Noise Generator, with the evolutionary search algorithm, Particle Swarm Optimization. The model is to adjust values of a trivial amount of pixels and significant reduce the accuracy of predictions of image classification neural networks, and so to protect the information contained in the original image data. The model is based on the principle that specific points on an original input image to classification neural networks show higher importance to final predict results. Also, the privacy protection model can inherit experiences from protection history of similar target images to the same target model and the same target image to other target models. The experiments evaluations on $ResNet50$ trained on $ImageNet$ dataset show that in practical black-box settings, the privacy protection model greatly alleviates the number of queries to the target model compared to the baseline algorithm with ensuring the success of protection. In the future, we aim to constructing a knowledge base for PSO of images in different categories and design a more intelligent initialization strategy for the PPNG model.

Acknowledgement. This research is supported, in part, by the SunTrust Fellowship Grant (ST20-07).

References

1. He, Z., Cai, Z., Yu, J.: Latent-data privacy preserving with customized data utility for social network data. IEEE Trans. Veh. Technol. **67**(1), 665–673 (2018)
2. Çiftçi, S., Akyüz, A.O., Ebrahimi, T.: A reliable and reversible image privacy protection based on false colors. IEEE Trans. Multimed. **20**(1), 68–81 (2018)
3. Cai, Z., He, Z.: Trading private range counting over big IoT data. In: 2019 IEEE 39th ICDCS, pp. 144–153, July 2019
4. Zhao, Y., Yu, Y., Li, Y., Han, G., Du, X.: Machine learning based privacy-preserving fair data trading in big data market. Inf. Sci. **478**, 449–460 (2019)
5. Gao, W., Yu, W., Liang, F., Hatcher, W.G., Lu, C.: Privacy-preserving auction for big data trading using homomorphic encryption. IEEE Trans. Netw. Sci. Eng., 1 (2018)
6. Abadi, M., et al.: Deep learning with differential privacy. In: Weippl, E.R., Katzenbeisser, S., Kruegel, C., Myers, A.C., Halevi, S. (eds.) ACM CCS, pp. 308–318. ACM (2016)
7. Shokri, R., Shmatikov, V.: Privacy-preserving deep learning. In: Ray, I., Li, N., Kruegel, C. (eds.) ACM CCS, pp. 1310–1321. ACM (2015)
8. Krizhevsky, A., Sutskever, I., Hinton, G.E.: ImageNet classification with deep convolutional neural networks. In: Pereira, F., Burges, C.J.C., Bottou, L., Weinberger, K.Q. (eds.) Advances in Neural Information Processing Systems 25. Curran Associates, Inc., pp. 1097–1105 (2012)
9. Zheng, X., Cai, Z., Li, Y.: Data linkage in smart internet of things systems: a consideration from a privacy perspective. IEEE Commun. Mag. **56**(9), 55–61 (2018)
10. https://aws.amazon.com/sagemaker/. Accessed 31 Jan 2020
11. https://cloud.google.com/mlengine/docs/technical-overview/. Accessed 31 Jan 2020
12. https://azure.microsoft.com/enus/services/machine-learning-studio/. Accessed 31 Jan 2020
13. Song, C., Ristenpart, T., Shmatikov, V.: Machine learning models that remember too much. CoRR abs/1709.07886 (2017)
14. Su, J., Vargas, D.V., Sakurai, K.: One pixel attack for fooling deep neural networks. IEEE Trans. Evol. Comput. **23**(5), 828–841 (2019)
15. Kennedy, J., Eberhart, R.C.: Particle swarm optimization. In: Proceedings of IEEE IJCNN, Washington, DC, USA, vol. 4, pp. 1942–1948. IEEE Computer Society, November 1995
16. Szegedy, C., et al.: Intriguing properties of neural networks. In: Bengio, Y., LeCun, Y. (eds.) ICLR (Poster) (2014)
17. Nguyen, A.M., Yosinski, J., Clune, J.: Deep neural networks are easily fooled: high confidence predictions for unrecognizable images. In: CVPR, pp. 427–436 IEEE Computer Society (2015)
18. Papernot, N., et al.: The limitations of deep learning in adversarial settings. In: EuroS&P, pp. 372–387. IEEE (2016)
19. Papernot, N., McDaniel, P.D., Goodfellow, I.J., Jha, S., Celik, Z.B., Swami, A.: Practical black-box attacks against machine learning. In: Karri, R., Sinanoglu, O., Sadeghi, A.R., Yi, X. (eds.) AsiaCCS, pp. 506–519. ACM (2017)
20. LeCun, Y., Cortes, C.: MNIST handwritten digit database (2010)
21. Stallkamp, J., Schlipsing, M., Salmen, J., Igel, C.: Man vs. computer: benchmarking machine learning algorithms for traffic sign recognition. Neural Netw. **32**, 323–332 (2012)

22. Krizhevsky, A.: Learning multiple layers of features from tiny images, pp. 32–33 (2009)
23. Zhang, Q., Wang, K., Zhang, W., Hu, J.: Attacking black-box image classifiers with particle swarm optimization. IEEE Access **7**, 158051–158063 (2019)
24. Mosli, R., Wright, M., Yuan, B., Pan, Y.: They might not be giants: crafting black-box adversarial examples with fewer queries using particle swarm optimization. CoRR abs/1909.07490 (2019)
25. He, K., Zhang, X., Ren, S., Sun, J.: Deep residual learning for image recognition. CoRR abs/1512.03385 (2015)
26. Deng, J., Socher, R., Fei-Fei, L., Dong, W., Li, K., Li, L.J.: ImageNet: a large-scale hierarchical image database. In: 2009 IEEE CVPR, vol. 00, pp. 248–255, June 2009
27. Simonyan, K., Zisserman, A.: Very deep convolutional networks for large-scale image recognition. In: ICLR (2015)

On Burial Depth of Underground Antenna in Soil Horizons for Decision Agriculture

Abdul Salam$^{(\boxtimes)}$ and Usman Raza

Department of Computer and Information Technology,
Purdue University, West Lafayette, IN 47906, USA
{salama,uraza}@purdue.edu

Abstract. Decision agriculture is the practice of accurately capturing the changing parameters of the soil including water infiltration and retention, nutrients supply, acidity, and other time changing phenomena by using the modern technologies. Using decision agriculture, fields can be irrigated more efficiently hence conserving water resources and increasing productivity. The Internet of Underground Things (IOUT) is being used to monitor the soil for smart irrigation. Moreover, the communication in wireless underground sensor networks is affected by soil characteristics such as soil texture, volumetric water content (VWC) and bulk density. These soil characteristics vary with soil type and soil horizons within a field. In this paper, we have investigated the effects of these characteristics by considering Holdrege soil series and homogeneous soil. It is shown that the consideration of soil characteristics of different soil horizons leads to 6% improved communication in wireless underground communications for smart agricultural practices.

Keywords: Cyber-physical systems · Underground electromagnetic propagation · Wireless underground sensor networks · Decision agriculture · Internet of things

1 Introduction

In decision agriculture, the soil horizons are the layers of soil which are formed by four soil processes and have unique chemical, physical, and visible characteristics. These soil process are additions, losses, transformations, and translocations. There are five horizons: O, A, E, B, and C. In soil, these horizons can form in any order. Some soils do not contain all horizons and in some soils multiple horizons can repeat. The horizon A and B are of most interest because of their high impact on plant growth. In wireless underground sensor networks, sensor nodes are buried in soil. Establishment of wireless communication links is important for data communication. As each soil horizon have unique soil texture, bulk density and water holding capability. Also depth and width of each horizon differs in different type of soils. These factors have a significant influence on the performance of a buried antenna and communication. These are (Fig. 1):

© Springer Nature Switzerland AG 2020
W. Song et al. (Eds.): ICIOT 2020, LNCS 12405, pp. 17–31, 2020.
https://doi.org/10.1007/978-3-030-59615-6_2

Fig. 1. The holdredge soil profile

Soil Moisture

Soil moisture changes with time due to climate and irrigation, which influence the soil permittivity.

Soil permittivity

Electromagnetic waves propagation in soil exhibit different characteristics in soil due to higher permittivity of soil.

Soil-Air Interface

Impedance of under ground antenna is changed because of current disturbance at antenna due to reflection from soil-air interface [18,41,52,54].

In this paper, by using the model for underground to underground (UG2UG) communications model, we have analyzed the performance of wireless underground channel by using Holdrege soil profile [57] and homogeneous soil. Moreover, we provide analytical results for path loss for three different scenarios including same soil moisture level across all horizons, water infiltration, and water retention scenario. Based on the analysis it is shown that antennas burried into soil horizons by taking soil characteristic into account experience less path loss as compared to antenna berried in homogeneous soil and path loss is decreased from 5–6 dB. It is also shown that path loss varies with soil moisture and increase in soil moisture also increase the path loss for all type of soils. It is also evident that in underground wireless sensor networks path loss increase with frequency therefore low operation frequencies are suitable for for wireless underground communication [15,19–50,53,59,60].

The rest of the paper is organized as follows: In Sect. 2, related work on communication in medium and the impact of the medium on antenna impedance is introduced. Section 3 gives the brief overview of soil properties. The impedance and the return loss of dipole antenna buried in soil are analyzed both theoretically and using simulations in Sect. 4, where an antenna impedance model considering the impact of the soil-air interface is developed. The experiment results are shown and analyzed in Sect. 5. Conclusions are drawn in Sect. 6.

We have used 31% sand particles and 29 clay particles soil for comparison with Holdrege soil.

Table 1. Physical properties of holdrege soil

Horizon	Depth in inches	Sand	Slit	Clay	Textual class
Ap	0–7	16.6	61.4	22.0	Silt loam
A	7–13	12.0	58.4	29.6	Silt clay loam
Bt1	13–16	13.3	55.3	31.4	Silt clay loam
Bt2	16–24	11.2	58.9	29.9	Silt clay loam

2   Related Work

Antennas used in WUSNs are buried in soil, which is uncommon in traditional communication scenarios. Antennas in matter have been analyzed in [14] where the electromagnetic fields of antennas in infinite dissipative medium and half space have been derived theoretically. In this analysis, the dipole antennas are assumed to be perfectly matched and hence the return loss is not considered. In [11], the impedance of a dipole antenna in solutions are measured. The impacts of the depth of the antenna with respect to the solution surface, the length of the dipole, and the complex permittivity of the solution are discussed. However, this work cannot be directly applied to WUSNs since the permittivity of soil has different characteristics than solutions and the change in the permittivity caused by the variation in soil moisture is not considered. The impact of these soil factors in underground communication has been analyzed in [19, 21–29, 32, 42–44].

In existing WUSN experiments and applications, the permittivity of the soil is generally calculated according to a soil dielectric model [1, 16], which leads to the actual wavelength of a given frequency. The antenna is then designed corresponding to the calculated wavelength [54]. Unfortunately, this approach often does not produce the desired antenna for the underground communication since the impedance of the antenna is not solely related to the wavelength of electromagnetic waves. In [54], an elliptical planar antenna is designed for a WUSN application. The size of the antenna is determined by comparing the wavelength in soil and the wavelength in air for the same frequency. However, this technique does not provide the desired impedance match. Moreover, when antennas are buried near the interface of a half-space, the impedance depends not only on the medium, but also on the reflections from the interface. This phenomenon is mentioned in [10], however, its impact is not modeled.

The disturbance caused by the interface is similar to the impedance change of a handheld device close to a human body [2, 55] or implanted devices in human body [3, 9]. In these applications, simulation and test bed results show that there are impacts from human body that cause performance degradation of the antennas. Though similar, these studies cannot be applied to the underground communication directly. First, the permittivity of the human body is higher than in soil. At 900 MHz, the relative permittivity of the human body is 50 [55] and for soil with a soil moisture of 5%, it is 5 [16]. In addition, the permittivity of soil varies with moisture, but for human body, it is relatively static.

Most importantly, in these applications, the human body can be modeled as a block while in underground communications, soil is modeled as a half-space since the size of the field is significantly larger than the antenna.

3 Soil Characteristics

We have used Holdredge soil and homogenoius soil for our analysis. Table 1 shows physical properties Holdrege soil [56]. We have selected Holdrege series because it is one of the well-drained, highly productive and most fertile soil in the Nebraska, United States. It is also official state soil of Nebraska and almost all the soil is under cultivation. As per United States Department of Agriculture [57]:

The Holdrege series of the soil is composed of in-depth, good drainage, mildly penetrable particles developed in calcium carbonate sediments. These highland soil contains sloppy areas which range form 0–15% with annual average temperature of approximately 55°F, and average annual rainfall is approximately at a particular location. It has fine particles of silt that are mixed with hyperactive, moist Typic Argiustolls.

Soils in the Holdrege series are recognized by features of their profile (created by horizontal layers) that are the result of the prairie environment. They are suggestive of soils formed under mixed grasses, in a climate where moisture stress is common, but where enough movement of water through the profile has resulted in downward movement of clays and lime. These processes have led to a soil with a thick, dark colored topsoil, a clay enriched subsoil and a substratum that contains free lime. Holdrege soils are among the most extensively cultivated soils in the state. Presently, nearly all Holdrege soils are cultivated. A very large part is irrigated. Corn and grain sorghum are the principal row crops. Winter wheat is the most commonly grown small grain. Their natural fertility, desirable tilth, and the landscape on which they exist join with irrigation water and the skillful management of Nebraska farms to provide a valuable agricultural resource [56].

4 Relative Permittivity of Soil

The EM wave propagation in soil is different from that of in air because of higher permittivity values of soil than that of air. Various soil factors effect the EM waves. These factors includes: soil texture, bulk density, soil moisture (also known as Volumetric Water Content), temperature and salinity. Relative permittivity has been investigated in detail by [5,16]. They define relative permittivity of various constituents (air, soil, bound and free water) of soil-water solution [5]. In [16], a semi-empirical permittivity model is presented which is used in this paper to find the effective permittivity of the soil-water mixture. Finally, the effective permittivity is calculated using the permittivity of all components, i.e., soil, water, and air, of the mixture.

4.1   The Impact of Soil on the Return Loss of an Antenna

Soil permittivity has direct effect on the return loss of an antenna. Variations in soil moisture causes the change in soil permittivity. This effect is visible in Fig. 3 which plots the effect of soil moisture on return loss of 70 mm monopole antenna. It can be observed that resonant frequency shifts to lower spectrum when the soil moisture is increased. An important thing to note is that return loss is minimum at resonant frequency f_{res}.

The primary reason of return loss is the impedance mismatch between the antennas, hence, it is important to calculate the impedance of the antenna. There is no closed form representation of antenna impedance, hence, impedance approximation given in [13] is used. This approximation is done for dipole antenna. Some other impedance approximation for dipole antennas are also presented in [14,61]. As per [13], impedance of dipole can be calculated as follow by using the induce-emf method [19]:

$$Imp_D \approx f_1(\gamma l_D) - i \left(120 \left(\ln \frac{2l_D}{D_D} - 1 \right) \cot(\gamma l_D) - f_2(\gamma l_D) \right), \qquad (1)$$

where

$$f_1(\gamma l_D) = -0.4787 + 7.3246\gamma l_D + 0.3963(\gamma l_D)^2 + 15.6131(\gamma l_D)^3 , \qquad (2)$$

$$f_2(\gamma l_D) = -0.4456 + 17.0082\gamma l_D - 8.6793(\gamma l_D)^2 + 9.6031(\gamma l_D)^3, \qquad (3)$$

where real portion of the wave number is given as γ, diameter of the dipole antenna is represented by D_D, and length (half) of the dipole is given by l. γl_D is calculated as follow:

$$\gamma l_D = \frac{2\pi l}{\lambda_{air}} \mathrm{Re}\left\{ \sqrt{\epsilon_s} \right\}, \qquad (4)$$

where subscript D represents the dipole antenna λ_{air} represents wavelength in air and ϵ_s represents the relative permittivity of the soil [16].

Soil permittivity ϵ_s rely on the frequency, therefore, γl_D and l_D/λ are not linearly related. Hence, when the antennas are deployed in soil instead of air, their impedance values (at resonant frequency) also becomes dependent on soil properties. This impedance mismatch due to different medium causes an antenna return loss which is given in dB as [19]:

$$RL_{dB} = 20 \log_{10} \left| \frac{Imp_{soil} + Imp_{air}}{Imp_{soil} - Imp_{air}} \right|. \qquad (5)$$

4.2   The Impact of Soil on Bandwidth

Bandwidth is also one of the major performance metric of the system. Shannon's equation [17] relates bandwidth of the system with channel capacity of medium. Shannon's equation shows that capacity is directly proportional to the bandwidth of the system. For wireless communications, antenna (return loss) also plays an important role in determining the final bandwidth of the system.

It has already been established in Sect. 4.1 that antenna return loss depends upon the frequency f and can be represented as $RL = R(f)$ and negative of return loss $-R(f)$ is given as S_{11}. For antenna operating at resonant frequency, bandwidth is given as the spectrum for which Δ values is higher than the negative of return loss. For all other operational frequencies, i.e., apart from resonant frequency, bandwidth will be less than resonant frequency. Following equation calculates the systems bandwidth for any operation frequency [7]:

$$B_{sys} = \begin{cases} 0 & \text{if } S_{11} > \Delta, \\ 2(f - f_{min}) & \text{if } S_{11} \leq \Delta \text{ and } f < f_{res}, \\ 2(f_{max} - f) & \text{if } S_{11} \leq \Delta \text{ and } f \geq f_{res}, \end{cases} \quad (6)$$

In above equation, resonant frequency is given by f_{res}, and f_{min} and f_{max} represents the minimum and maximum frequencies, respectively, for which $R(f) \leq \Delta$.

As an example for estimation of antenna bandwidth, S_{11} is plotted with f. The operating frequency of the antenna is 24 MHz less than the values for resonant frequency and $\Delta = -10$ dB. The bandwidth is calculated as 14 MHz, S_{11} remains lower than Δ for whole spectral band.

4.3   The Impact of Soil on Path Loss

A detailed investigation is performed in [6,8] to understand the communication in WUSNs. The effect of soil on aboveground-to-underground (UG2AG) & underground-to-aboveground (AG2UG) channel has been studied in detail. It was found that EM waves attenuation in the soil is dependent on various factors such as: distance, soil moisture, and soil type. Irrespective of the direction, total path loss PL_T is calculated as:

$$PL_T = \left(PL_{ug}(d_{ug}) + PL_{ag}(d_{ag}) + PL_{(R, \rightarrow)} \right), \quad (7)$$

In above equation, losses in both aboveground & underground area is given by $PL_{ag}(d_{ag})$ and $PL_{ug}(d_{ug})$, respectively. Moreover, depending upon the direction of the wave propagation $\rightarrow$, $PL_{(R, \rightarrow)}$ gives the refraction loss. The direction could be either $ag2ug$ or $ug2ag$.

The losses in Eq. (7) for both UG and AG environment are calculated as [58]:

$$PL_{ug}(d_{ug}) = 6.4 + 20 \log d_{ug} + 20 \log \gamma + 8.69 \alpha_{(const, soil)} d_{ug}, \quad (8)$$
$$PL_{ag}(d_{ag}) = -147.6 + 10 \alpha_{(coef, air)} \log d_{ag} + 20 \log f, \quad (9)$$

In above equation, the terms $\alpha_{(coef, air)}$ represents the attenuation coefficient in air, $\alpha_{(const, soil)}$ represents the attenuation constant, f represent the operation frequency and γ gives the phase shifting constant. The $\alpha_{(coef, air)} > 2$ because of ground reflection effect. The empirical experiments in [4] shows that $\alpha_{(coef, air)}$ values lies in the range of 2.8–3.3. In Eq. (8), $\alpha_{(const, soil)}$ and γ are used to incorporate the impact of soil on signal attenuation. The values for $\alpha_{(const,soil)}$ and γ is given as:

$$k_s = \alpha_{(const,soil)} + i\gamma = i\omega\sqrt{\mu_0\epsilon_s}, \tag{10}$$

In above equation, k_s, μ_0, and ϵ represents the soil propagation constant, free space permeability, and soil effective permittivity, respectively.

Owing to the higher values of soil permittivity, EM waves can only penetrate the soil-air interface, if the incident angle θ_t is small, and are reflected or refracted otherwise. Therefore, waves with small θ_t in are able to perform UG2AG propagation, and refracted angle $\rightarrow 0$ for AG2UG propagation. Moreover, AG2UG propagation is vertical in soil. Therefore, for AG2UG and UG2AG communication links, the underground distance traveled by the wave is approximated as the burial depth h_u, i.e., $d_{ug} \simeq h_{ug}$. Similarly, aboveground communication path is approximated using height of AG node h_{ag} and horizontal distance between both nodes $d_{ag \leftrightarrow ug}$. The aboveground path is given as: $d_{ag} = \sqrt{h_{ag}^2 + d_{ag \leftrightarrow ug}^2}$.

A maximum power path, i.e., where $\theta_i \rightarrow 0$, is considered for the AG2UG link. Therefore, approximation of refraction loss in Eq. (7) is given as follow [12]:

$$PL_{(R,ag2ug)} \simeq 20\log\frac{r_i + 1}{4}, \tag{11}$$

where refractive index of soil is represented by r_i. r_i is calculated in [8] as follow:

$$r_i = \sqrt{\frac{\sqrt{\epsilon'^2 + \epsilon''^2} + \epsilon'}{2}}. \tag{12}$$

Moreover, for UG2AG link. signal travels from the medium of high density to lower density, therefore, energy of the signal is refracted, i.e., $L_{(R,ug2ag)} = 0$.

4.4 Channel Capacity of Wireless Underground Communications

In addition to bandwidth, capacity of channel also effect the underground communication performance. To that end, the effect of soil properties on channel capacity is investigated. As per Shannon equation, capacity is dependent upon bandwidth B, noise N, and strength of the received signal R [7]:

$$C = B_{sys}\log_2(1 + \frac{R}{NB_{sys}}), \tag{13}$$

For this analysis, maximum achievable bandwidth is considered. As show in Eq. (6), this maximum bandwidth is estimated by antenna design. Antenna properties (return loss and path loss) will effect the power transmitted by the sender node P_t. Therefore, the received signal strength (dB) is calculated using antenna return loss in Eq. (5) and antenna path loss in Eq. (7). The received signal is given as [7] (Fig. 2):

$$R_{dB} = P_t + 10\log_{10}(1 - 10^{-\frac{RL_{dB}}{10}}) - PL_T, \tag{14}$$

Moreover, the above signal strength is based on discussion in Sect. 4.1 and Sect. 4.3.

Underground noise is stable during the testbeds experiments, hence, N can be used as a constant value [51].

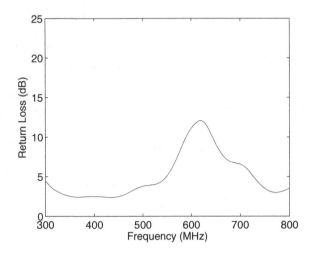

Fig. 2. Return loss of the antenna

5   Numerical Analysis

We have considered three cases for analytical evaluation. First case we have compared the two soils under the same soil moisture case for all soil horizons and depths. In second case we analyses the water infiltration scenario in which top soil horizons have more water content than the subsoil horizons. Third case compares the drainage scenario in which subsoil is more saturated as compared to the topsoil. We have used frequency range of 300 MHz to 800. Transmitted power is 15 dBm. Return Loss of the antenna used in the evaluation is shown in Figure. Antennas are buried at four depths. Four antenna burial depth corresponds to four different horizons (Ap, A, Bt1, Bt2) of Holdrege soil as shown in Table 1. For homogeneous soil these are 10 cm, 20 cm, 30 cm and 40 cm. Horizontal distance between transmitter receiver is 50 cm. Bulk density is $1.5\,\mathrm{g/cm^3}$ and particle density is $2.66\,\mathrm{g/cm^3}$.

5.1   Same Soil Moisture Scenario

Figure 3 shows the path loss for two soil types for Volumetric Water Content (VWC) of value of 10%. For all depths and across all frequency range Path loss for homogeneous soil is 5 dB to 6 dB higher than as compared to Holdrege soil.

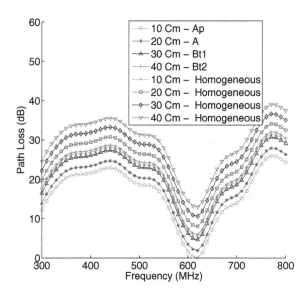

Fig. 3. Path loss vs. frequency - VWC 10%

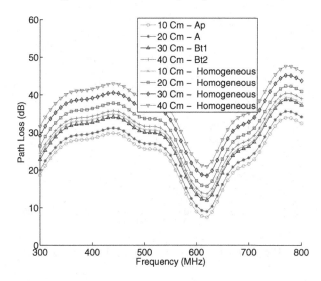

Fig. 4. Path loss vs. frequency - VWC 20%

Moreover between 550 MHz to 650 MHz range path loss is low because of the low return loss of the antenna. It is also clear that path loss increases with frequency.

Figure 4 shows the path loss for two soil types for Volumetric Water Content (VWC) of value of 20%. For all depths and across all frequency range Path loss for homogenous soil is 5 dB to 6 dB higher than as compared to Holdrege soil. Due to 10% increase in water content there is an increase of 8 dB for all horizons.

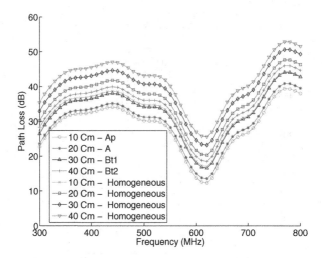

Fig. 5. Path loss vs. frequency - VWC 30%

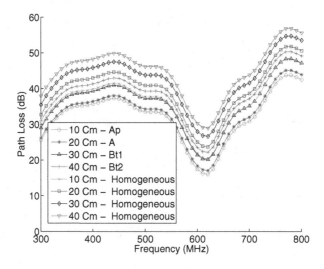

Fig. 6. Path loss vs. frequency - VWC 40%

Figure 5 and Fig. 6 shows the path loss for two soil types for Volumetric Water Content (VWC) of value of 30% and 40%. For both soil moisture levels, for all depths and across all frequency range path loss for homogenous soil is 5 dB to 6 dB increased as compared to Holdrege soil. Path loss for 30% and 40% is considerably higher than dry than the 10%.

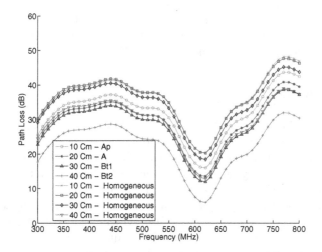

Fig. 7. Path loss vs. frequency - water infiltration scenario

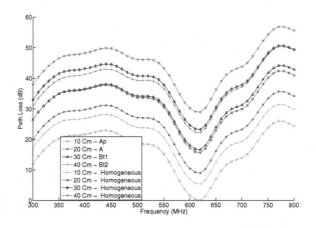

Fig. 8. Path loss vs. frequency - drainage scenario%

5.2   Water Infiltration Scenario

In this case we consider the scenario in which higher horizons have more water content as compared to lower soil horizons. Figure 7 shows the path loss when Ap horizon have 40% VWC, A horizon have 30% VWC, Bt1 have 20% VWC and Bt2 have 10% VWC. It is evident that communication performance is best at Bt2 horizon because of low water content.

5.3   Water Retention Scenario

In this case we consider the scenario in which lower horizons have more water content as compared to higher soil horizons. Figure 8 shows the path loss when

Ap horizon have 10% VWC, A horizon have 20% VWC, Bt1 have 30% VWC and Bt2 have 40% VWC. Antenna buried at the A horizon experience lower path loss because of low attenuation due to lower VWC.

6 Conclusions

In this paper, the impacts of soil texture, soil moisture on burial depth of antenna in different soil horizons and on path loss are analyzed for underground wireless communications in Holdrege soil and homogeneous soil. It is shown that antennas buried into soil horizons by taking soil characteristic into account experience less path loss as compared to antenna berried in homogeneous soil. It is also shown that path loss varies with soil moisture and increase in soil moisture also increase the path loss for all type of soils. It is also evident that in underground wireless sensor networks path loss increase with frequency therefore low operation frequencies are suitable for wireless underground communication.

References

1. Akyildiz, I.F., Sun, Z., Vuran, M.C.: Signal propagation techniques for wireless underground communication networks. Phys. Commun. J. **2**(3), 167–183 (2009)
2. Boyle, K., Yuan, Y., Ligthart, L.: Analysis of mobile phone antenna impedance variations with user proximity. IEEE Trans. Antennas Propag. **55**(2), 364–372 (2007)
3. Dissanayake, T., Esselle, K., Yuce, M.: Dielectric loaded impedance matching for wideband implanted antennas. IEEE Trans. Microw. Theory Tech. **57**(10), 2480–2487 (2009)
4. Do, T., Gan, L., Nguyen, N., Tran, T.: Fast and efficient compressive sensing using structurally random matrices. IEEE Trans. Signal Process. **60**(1), 139–154 (2012)
5. Dobson, M., Ulaby, F., Hallikainen, M., El-Rayes, M.: Microwave dielectric behavior of wet soil–Part II: dielectric mixing models. IEEE Trans. Geosci. Remote. Sens. **GE-23**(1), 35–46 (1985)
6. Dong, X., Vuran, M.C.: A channel model for wireless underground sensor networks using lateral waves. In: Proceedings of IEEE Globecom 2011, Houston, TX, December 2011
7. Dong, X., Vuran, M.: Impacts of soil moisture on cognitive radio underground networks. In: 2013 First International Black Sea Conference on Communications and Networking (BlackSeaCom), pp. 222–227, July 2013
8. Dong, X., Vuran, M.C., Irmak, S.: Autonomous precision agriculture through integration of wireless underground sensor networks with center pivot irrigation systems. Ad Hoc Netw. **11**, 1975–1987 (2012)
9. Gosalia, K., Humayun, M., Lazzi, G.: Impedance matching and implementation of planar space-filling dipoles as intraocular implanted antennas in a retinal prosthesis. IEEE Trans. Antennas Propag. **53**(8), 2365–2373 (2005)
10. Hunt, K., Niemeier, J., Kruger, A.: RF communications in underwater wireless sensor networks. In: IEEE International Conference on Electro/Information Technology (EIT), Normal, IL, May 2010

11. Iizuka, K.: An experimental investigation on the behavior of the dipole antenna near the interface between the conducting medium and free space. IEEE Trans. Antennas Propag. **12**(1), 27–35 (1964)
12. Johnk, C.T.: Engineering Electromagnetic Fields and Waves, 2nd edn. Wiley, Hoboken (1988)
13. Johnson, R.C. (ed.): Antenna Engineering Handbook, 3rd edn. McGraw-Hill Inc., New York (1993)
14. King, R.W.P., Smith, G.S.: Antennas in Matter. The MIT Press, Cambridge (1981)
15. Konda, A., et al.: Soft microreactors for the deposition of conductive metallic traces on planar, embossed, and curved surfaces. Adv. Funct. Mater. **28**(40), 1803020. https://onlinelibrary.wiley.com/doi/abs/10.1002/adfm.201803020
16. Peplinski, N., Ulaby, F., Dobson, M.: Dielectric properties of soil in the 0.3-1.3 GHz range. IEEE Trans. Geosci. Remote. Sens. **33**(3), 803–807 (1995)
17. Proakis, J., Salehi, M.: Digital Communications, 5th edn. McGraw-Hill, New York (2007)
18. Ritsema, C.J., Kuipers, H., Kleiboer, L., Elsen, E., Oostindie, K., Wesseling, J.G., Wolthuis, J., Havinga, P.: A new wireless underground network system for continuous monitoring of soil water contents. Water Resour. Res. J. **45**, 1–9 (2009)
19. Salam, A., Vuran, M.C., Dong, X., Argyropoulos, C., Irmak, S.: A theoretical model of underground dipole antennas for communications in internet of underground things. IEEE Trans. Antennas Propag. **67**(6), 3996–4009 (2019)
20. Salam, A., Vuran, M.C.: Impacts of soil type and moisture on the capacity of multi-carrier modulation in internet of underground things. In: Proceedings of the 25th ICCCN 2016, Waikoloa, Hawaii, USA, August 2016
21. Salam, A.: Pulses in the sand: long range and high data rate communication techniques for next generation wireless underground networks. ETD collection for University of Nebraska - Lincoln (AAI10826112) (2018). http://digitalcommons.unl.edu/dissertations/AAI10826112
22. Salam, A.: A comparison of path loss variations in soil using planar and dipole antennas. In: 2019 IEEE International Symposium on Antennas and Propagation. IEEE, July 2019
23. Salam, A.: Design of subsurface phased array antennas for digital agriculture applications. In: Proceedings of the 2019 IEEE International Symposium on Phased Array Systems and Technology (IEEE Array 2019), Waltham, MA, USA, October 2019
24. Salam, A.: A path loss model for through the soil wireless communications in digital agriculture. In: 2019 IEEE International Symposium on Antennas and Propagation. IEEE, July 2019
25. Salam, A.: Sensor-free underground soil sensing. In: ASA, CSSA and SSSA International Annual Meetings. ASA-CSSA-SSSA (2019)
26. Salam, A.: Subsurface MIMO: a beamforming design in internet of underground things for digital agriculture applications. J. Sens. Actuator Netw. **8(3)** (2019). https://www.mdpi.com/2224-2708/8/3/41
27. Salam, A.: Underground environment aware MIMO design using transmit and receive beamforming in internet of underground things. In: Issarny, V., Palanisamy, B., Zhang, L.-J. (eds.) ICIOT 2019. LNCS, vol. 11519, pp. 1–15. Springer, Cham (2019). https://doi.org/10.1007/978-3-030-23357-0_1
28. Salam, A.: An underground radio wave propagation prediction model for digital agriculture. Information **10**(4) (2019). http://www.mdpi.com/2078-2489/10/4/147

29. Salam, A.: Underground soil sensing using subsurface radio wave propagation. In: 5th Global Workshop on Proximal Soil Sensing. COLUMBIA, MO, May 2019

30. Salam, A.: Internet of things for environmental sustainability and climate change. In: Salam, A. (ed.) Internet of Things for Sustainable Community Development. IT, pp. 33–69. Springer, Cham (2020). https://doi.org/10.1007/978-3-030-35291-2_2

31. Salam, A.: Internet of things for sustainability: perspectives in privacy, cybersecurity, and future trends. In: Salam, A. (ed.) Internet of Things for Sustainable Community Development. IT, pp. 299–327. Springer, Cham (2020). https://doi.org/10.1007/978-3-030-35291-2_10

32. Salam, A.: Internet of Things for Sustainable Community Development, 1st edn. Springer, Cham (2020). https://doi.org/10.1007/978-3-030-35291-2

33. Salam, A.: Internet of things for sustainable community development: introduction and overview. In: Salam, A. (ed.) Internet of Things for Sustainable Community Development. IT, pp. 1–31. Springer, Cham (2020). https://doi.org/10.1007/978-3-030-35291-2_1

34. Salam, A.: Internet of things for sustainable forestry. In: Salam, A. (ed.) Internet of Things for Sustainable Community Development. IT, pp. 147–181. Springer, Cham (2020). https://doi.org/10.1007/978-3-030-35291-2_5

35. Salam, A.: Internet of things for sustainable human health. In: Salam, A. (ed.) Internet of Things for Sustainable Community Development. IT, pp. 217–242. Springer, Cham (2020). https://doi.org/10.1007/978-3-030-35291-2_7

36. Salam, A.: Internet of things for sustainable mining. In: Salam, A. (ed.) Internet of Things for Sustainable Community Development. IT, pp. 243–271. Springer, Cham (2020). https://doi.org/10.1007/978-3-030-35291-2_8

37. Salam, A.: Internet of things for water sustainability. In: Salam, A. (ed.) Internet of Things for Sustainable Community Development. IT, pp. 113–145. Springer, Cham (2020). https://doi.org/10.1007/978-3-030-35291-2_4

38. Salam, A.: Internet of things in agricultural innovation and security. In: Salam, A. (ed.) Internet of Things for Sustainable Community Development. IT, pp. 71–112. Springer, Cham (2020). https://doi.org/10.1007/978-3-030-35291-2_3

39. Salam, A.: Internet of things in sustainable energy systems. In: Salam, A.A. (ed.) Internet of Things for Sustainable Community Development. IT, pp. 183–216. Springer, Cham (2020). https://doi.org/10.1007/978-3-030-35291-2_6

40. Salam, A.: Internet of things in water management and treatment. In: Salam, A. (ed.) Internet of Things for Sustainable Community Development. IT, pp. 273–298. Springer, Cham (2020). https://doi.org/10.1007/978-3-030-35291-2_9

41. Salam, A.: Wireless underground communications in sewer and stormwater overflow monitoring: radio waves through soil and asphalt medium. Information 11(2), 98 (2020)

42. Salam, A., et al.: The future of emerging IoT paradigms: architectures and technologies (2019)

43. Salam, A., Karabiyik, U.: A cooperative overlay approach at the physical layer of cognitive radio for digital agriculture. In: Third International Balkan Conference on Communications and Networking 2019 (BalkanCom 2019), Skopje, Macedonia, the former Yugoslav Republic of, June 2019

44. Salam, A., Shah, S.: Internet of things in smart agriculture: enabling technologies. In: 2019 IEEE 5th World Forum on Internet of Things (WF-IoT) (WF-IoT 2019), Limerick, Ireland, April 2019

45. Salam, A., Vuran, M.C.: EM-based wireless underground sensor networks. In: Pamukcu, S., Cheng, L. (eds.) Underground Sensing, pp. 247–285. Academic Press, Cambridge (2018)

46. Salam, A., Vuran, M.C., Irmak, S.: Di-sense: in situ real-time permittivity estimation and soil moisture sensing using wireless underground communications. Comput. Netw. **151**, 31–41 (2019). http://www.sciencedirect.com/science/article/pii/S1389128618303141

47. Salam, A., Vuran, M.C.: Smart underground antenna arrays: a soil moisture adaptive beamforming approach. In: Proceedings of the IEEE INFOCOM 2017, Atlanta, USA, May 2017

48. Salam, A., Vuran, M.C.: Wireless underground channel diversity reception with multiple antennas for internet of underground things. In: Proceedings of the IEEE ICC 2017, Paris, France, May 2017

49. Salam, A., Vuran, M.C., Irmak, S.: Pulses in the sand: Impulse response analysis of wireless underground channel. In: The 35th Annual IEEE International Conference on Computer Communications (INFOCOM 2016), San Francisco, USA, April 2016

50. Salam, A., Vuran, M.C., Irmak, S.: Towards internet of underground things in smart lighting: a statistical model of wireless underground channel. In: Proceedings of the 14th IEEE International Conference on Networking, Sensing and Control (IEEE ICNSC), Calabria, Italy, May 2017

51. Silva, A.R., Vuran, M.C.: Empirical evaluation of wireless underground-to-underground communication in wireless underground sensor networks. In: Proceedings of IEEE International Conference on Distributed Computing in Sensor Systems (DCOSS 2009), pp. 231–244, Marina del Rey, CA, June 2009

52. Silva, A.R., Vuran, M.C.: (CPS)^2: integration of center pivot systems with wireless underground sensor networks for autonomous precision agriculture. In: Proceedings of ACM/IEEE International Conference on Cyber-Physical Systems, Stockholm, Sweden, pp. 79–88, April 2010

53. Temel, S., Vuran, M.C., Lunar, M.M., Zhao, Z., Salam, A., Faller, R.K., Stolle, C.: Vehicle-to-barrier communication during real-world vehicle crash tests. Comput. Commun. **127**, 172–186 (2018). http://www.sciencedirect.com/science/article/pii/S0140366417305224

54. Tiusanen, M.J.: Wireless Soil Scout prototype radio signal reception compared to the attenuation model. Precis. Agric. **10**(5), 372–381 (2008)

55. Toftgard, J., Hornsleth, S., Andersen, J.: Effects on portable antennas of the presence of a person. IEEE Trans. Antennas Propag. **41**(6), 739–746 (1993)

56. UNL Soil Website. http://snr.unl.edu/data/publications/HoldregeSoil.asp#sand. Accessed Jan 2020

57. USDA Website. https://soilseries.sc.egov.usda.gov/OSD_Docs/H/HOLDREGE.html. Accessed Jan 2020

58. Vuran, M.C., Akyildiz, I.F.: Channel model and analysis for wireless underground sensor networks in soil medium. Phys. Commun. **3**(4), 245–254 (2010)

59. Vuran, M.C., Salam, A., Wong, R., Irmak, S.: Internet of underground things in precision agriculture: architecture and technology aspects. Ad Hoc Netw. (2018). http://www.sciencedirect.com/science/article/pii/S1570870518305067

60. Vuran, M.C., Salam, A., Wong, R., Irmak, S.: Internet of underground things: Sensing and communications on the field for precision agriculture. In: 2018 IEEE 4th World Forum on Internet of Things (WF-IoT) (WF-IoT 2018), Singapore, February 2018

61. Wu, T.: Theory of the dipole antenna and the two-wire transmission line. J. Math. Phys. **2**, 550–574 (1961)

Deriving Interpretable Rules for IoT Discovery Through Attention

Franck Le[(✉)] and Mudhakar Srivatsa

IBM T.J. Watson Research Center, Yorktown Heights, USA
{fle,msrivats}@us.ibm.com

Abstract. Due to their high vulnerability, IoT has become a primary target for cybercriminals (e.g., botnets, network infiltration). As a result, many solutions have been developed to help users and administrators identify IoT devices. While solutions based on deep learning have been shown to outperform traditional approaches in other domains, their lack of explanation and their inference latency present major obstacles for their adoption in network traffic analysis, where throughputs of Gbps are typically expected. Extracting rules from a trained neural network presents a compelling solution, but existing methods are limited to feed-forward networks, and RNN/LSTM. In contrast, attention-based models are a more recent architecture, and are replacing RNN/LSTM due to their higher performance. In this paper, we therefore propose a novel efficient algorithm to extract rules from a trained attention-based model. Evaluations on actual packet traces of more than 100 IoT devices demonstrate that the proposed algorithm reduces the storage requirements and inference latency by 4 orders of magnitude while still achieving an average f1-score of 0.995 and a fidelity score of 98.94%. Further evaluation on an independent dataset also shows improved generalization performance: The extracted rules achieve better performance, especially thanks to their inherent capability to identify unknown devices.

1 Introduction

Internet of Things (IoT) – the extension of Internet connectivity to objects (e.g., from smart electronic appliances to industrial sensors) beyond traditional devices (e.g., desktops, laptops, smartphones or tablets) – introduces new major challenges to IT departments. More specifically, with their widespread adoption and poor security [1–3], IoT devices constitute a prime target for attacks, e.g., as ingress points to broader IT infrastructure. For these reasons, a number of approaches have recently been developed to discover them through passive network monitoring (e.g., [4–11]). Among the existing solutions, those based on deep learning can achieve high performance (e.g., accuracy, f1-score) but present limitations that can limit their adoption in operational environments: First, the inference latency of deep models can be significantly larger than those of pattern matching based rules traditionally used in Network Intrusion Detection Systems (NIDS) [12,13]. For example, our internal experiments show that the

© Springer Nature Switzerland AG 2020
W. Song et al. (Eds.): ICIOT 2020, LNCS 12405, pp. 32–44, 2020.
https://doi.org/10.1007/978-3-030-59615-6_3

inference latency of feedforward networks, Long Short-Term Memory (LSTM), and Bayesian Neural Networks can be 4 to 6 orders of magnitude larger than those of pattern matching based rules. These inference latency can consequently increase the packet processing time which is critical to achieve the required high throughput for network analysis. Second, deep models are notoriously difficult to interpret. Operating as "black-boxes", they do not provide any explanation behind their outcomes. Considerable research efforts have been devoted to address this limitation, and towards extracting the knowledge stored in a network's weights through concise symbolic forms such as Boolean and fuzzy if-then clauses, or finite-state automata (e.g., [14–20]), but existing solutions mainly apply to feedforward networks, and RNN/LSTMs.

However, recently, attention-based models [21] have been replacing RNNs and LSTMs due to their higher performance. Attention is considered one of the most influential ideas in Deep Learning, and is the current state-of-the-art for sequence learning. In this paper, we therefore present a novel rule extraction algorithm for and from trained attention-based models. To the best of the authors' knowledge, there is currently no algorithm to extract rules from these reigning architectures. The proposed algorithm is efficient as it builds upon the attention weights readily made available by the attention-based models. In addition, by extracting the knowledge in the form of Boolean clauses, the rules not only provide explanation capability, but also can be directly implemented in current NIDS frameworks at high speeds, and offer improved generalization performance.

We evaluate the proposed algorithm on real network traces from hundreds of IoT devices captured over an entire year. More specifically, we trained several attention-based models to discover the IoT devices from their network traffic, and infer their properties, before applying our proposed algorithm to extract the knowledge in Boolean if-then clauses. The results show that the extracted rules not only reduce the storage requirements and inference latency by 4 orders of magnitude, but also achieve an average f1-score of 0.995 in comparison to the initial average f1-score of 0.998 from the attention-based model. In other words, the rules offer interpretability with a minimal loss in f1-score. We provide examples of the rules learned through *Alexandr*, to demonstrate the interpretability of the rules. The rules also achieve a fidelity score of 98.94%. The fidelity metric represents how accurately the extracted rules correspond to the initial trained attention-based model. Finally, we evaluate the rules on an independent dataset, and show that the rules achieve improved generalized performance compared to the attention-based model. More specifically, while the attention-based model achieves an average f1-score of 0.553 on the new dataset, the rules achieve an f1-score of 0.728, especially thanks to their inherent capability to identify unknown devices (i.e., instances not matching any rule).

2 Background: Attention-Based Model

Attention [21] was initially designed for Neural Machine Translation, and has since been adopted in other fields including image captioning [22], and speech recognition [23].

Before the introduction of attention, the state-of-the-art in neural machine translation at that time relied on an encoder-decoder architecture [24,25], where the encoder compresses the input sequence (i.e., sentence to be translated) into a vector of a fixed length; and from which, the decoder then generates the output sequence (i.e., translation sentence). Both encoder and decoder typically consist of recurrent neural network (RNN), and while this approach worked well for short sequences, the performance dropped with longer sentences due to the inability to compress very long sequences into a fixed-length vector. The attention mechanism was designed to address this problem.

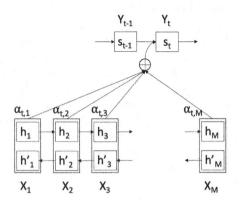

Fig. 1. Illustration of the attention weights

Rather than encoding the input sequence into a fixed-length vector, attention-based methods work by looking at specific parts of the input sequence in order to generate each token of the output sequence. In other words, attention-based models pay more attention (hence, the name) to different parts in the input sentence, to generate each output token. In order to achieve it, a RNN (e.g.., bidirectional RNN) is first applied to the input sequence. The intermediate states of the RNN provide a rich set of features for each token of the input sentence. The output sequence is then generated through another RNN which as part of its input takes the previously computed intermediate states, weighted by a set of attention weights for each output token.

Figure 1 (extracted from [21][1]) illsutrates the main concepts: $(X_1, X_2, ..., X_M)$ is the input sequence. $(h_1, h_2, ..., h_M)$ represents the forward hidden states of the first (bidirectional) RNN. $(h'_1, h'_2, ..., h'_M)$ designates the corresponding backward hidden states. H_i composed of the concatenation of h_i, and h'_i, provides a summary of X_i, and the tokens surrounding it. The target output token Y_t is generated by a second RNN. s_t is the hidden state for time t. It is computed taking into account not only the previous hidden state s_{t-1}, the previous output token Y_{t-1} but also the context $\sum_{j=1}^{M} \alpha_{tj} H_j$. The weights α_{tj} represent the

---

[1] Notations were slightly modified.

attention weights. They reflect which input tokens are important for each output token, and are learned through Gradient Descent and Back-propagation. For further details on attention-based models, we refer readers to [21].

3   Alexandr Algorithm

3.1   What Is the Central Idea?

This section describes *Alexandr*, a novel algorithm for extracting rules from a trained attention-based model. The algorithm builds upon the attention weights which by design, are learned to identify for each token of the output sequence, the important tokens from the input sequence (Sect. 2). When adopting an attention-based model as a classifier, the weights therefore indicate the important tokens that lead to the prediction outcome. Finally, by aggregating over all the input sequences that are classified by the attention-based model into a given class, the attention weights can reveal the relevant sequence patterns for that class.

To illustrate the intuition, we consider the following sequence of DNS queries in Fig. 2 that is submitted by an IoT device: *device-metrics-us.amazon.com, ntp.amazon.com, ntp.amazon.com,.., pindorama.amazon.com.* From such sequence, an attention-based model is trained to predict the *vendor* of the IoT device. After vulnerabilities in devices from Belkin, Shenzhen, Ring and other vendors were revealed and shown to be exploitable by hackers [26], admins may want to know if any IoT device from those vendor is present in their network. The trained attention model provides not only a prediction for the vendor (e.g., *amazon*), but also as illustrated in Fig. 2, the attention weights which indicate the important (sub-sequence of) DNS queries.

device-metrics-us.amazon.com	ntp.amazon.com	ntp.amazon.com	...	pindorama.amazon.com
3e-06	0.12	0.19	...	1e-08

Fig. 2. Illustration of attention weights. The first row represents the sequence of queried domains, and the second row indicates the attention weight associated with each domain by the attention-based model.

3.2   How Are the Rules Extracted?

We assume an attention-based model m, trained to classify input sequences from $\mathcal{X}$ into classes in $\mathcal{C}$, and a dataset $\mathcal{D} = (x_1, x_2, ..., x_M)$ comprising of M instances in $\mathcal{X}$. Each instance x_i ($i \in [1, M]$) is a sequence of up to L tokens. Algorithm 1 presents the main steps. For ease of illustration, the procedure describes the steps to extract the rule for one given class c. However, the code can be easily extended to generate the rules for all the classes in one single pass of the dataset $\mathcal{D}$.

First, for each instance x from $\mathcal{D}$, m is applied to x, with y denoting the predicted class, and *attention* representing the attention weights for each of the

Algorithm 1. *Alexandr* algorithm

```
 1: procedure ALEXANDR(m, D, c)
 2:     for x in D do
 3:         y, attention ← m.predict(x)
 4:         if y == c then
 5:             for window.size in [1, max_size]  do
 6:                 while slide window over x do
 7:                     pattern ← x[window]
 8:                     Sum weights of tokens in pattern
 9:                     Update weights for pattern
10:                 end while
11:             end for
12:         end if
13:     end for
14:     Sort pattern in decreasing values of weights
15:     Extract top max_len patterns into list
16:     for k in range(max_len): do
17:         Generate all disjunctions of length k from list
18:                                      ▷ Each combination is a candidate rule
19:         Evaluate candidate rule over D
20:         Compute f1-score of candidate rule
21:     end for
22:     Return candidate rule with highest f1-score
23: end procedure
```

tokens in x (line 3). If y equals the target class c (line 4), then we vary the size of a window, and slide the window over x (lines 5 & 6). The tokens in the window constitute a sequence pattern (line 7). The sum of the attention weights corresponding to the tokens in the sequence pattern is calculated (line 8), and the weight of the sequence pattern is then updated (line 9.)

To illustrate steps 5 from 9, we continue with the example from Fig. 2. For a window of size 2, the window would first be set to "*device-metrics-us.amazon.com, ntp.amazon.com*", and the weight of this pattern would be set to (3e-06+0.12). At the next step, the window would be set to "*ntp.amazon.com, ntp.amazon.com*" (since the window size is 2), and the weight of that sequence to (0.12+0.19).

After iterating over all instances from D, we obtain a set of patterns, each associated with a weight. We then generate all combinations of disjunctions of up to the top *max_len* patterns. Each combination represents a candidate rule. We evaluate each candidate rule over D, and returns the one which achieves the largest f1-score (or desired metric).

4 Evaluation

4.1 Dataset

We evaluate *Alexandr* on two independent datasets. The first one, dataset 1, consists of the network traffic from 108 IoT devices deployed in a private lab. The traffic is captured between January 1 2016, and December 31 2017. The second dataset, dataset 2, is the publicly available packet trace captured at UNSW from 21 IoT devices, from September 22, 2016 to October 11, 2016 [4].

For each dataset, we focus on the DNS traffic. More specifically, we define an instance as the sequence of DNS requests submitted by an IoT device over a period of a day. Each instance is labelled with the vendor of the IoT device. Because devices are continuously added and removed from the lab, the number of active devices on each day varies, and dataset 1 comprises 10,915 instances from 41 vendors (Fig. 3 provides an excerpt of the IoT devices in dataset 1), while dataset 2 includes 341 instances from 14 vendors. Eleven vendors (*amazon, belkin, google, ihome, lifx, netatmo, pixstar, samsung, smartlabs, tplink, withings,*) are present in both datasets.

Amazon Echo	Honeywell Thermostat
Amazon FireTV	iHome Plug
Apple TV	iRobot Roomba
Arcsoft Camera	LG TV
Belkin Camera	Lifx Bulb
Belkin Motion-sensor	Logitech Harmony
Belkin Plug	Makerbot 3D Printer
Belkin Wemolink	Movo Camera
Belkin Wemoswitch	Netatmo Camera
Canary Camera	Netatmo Weather-station
D-Link Camera	Netgear Arlohub
Ecobee Thermostat	Philips Bridge
Foscam Camera	Pixstar Fotoconnect
Google Camera	Ring Doorbell
Google Chromecast	Samsung Hub
Google Nexus	Samsung Refrigerator
Google Revue	Skybell Doorbell
Google Thermostat	TP-Link Hub
Gracedigital Mondo	Withings Aura

Fig. 3. Excerpt of IoT devices in Dataset 1

4.2 Methodology

The goals are to train an attention-based model, and extract the rules using *Alexandr* using one dataset; and then to emulate a deployment-in-the-wild scenario by comparing the results from the attention-based model and the extracted rules on the other dataset. Because dataset 1 is larger than dataset 2, both in the number of instances, and number of classes, we train the attention-based model, and extract rules using *Alexandr* on dataset 1, and then use dataset 2 to understand how well the results generalize.

- **Preprocessing:** We preprocess the data by discarding DNS queries which are not conducive to the classification objective, i.e., predicting the vendor of the IoT device. This is similar to the elimination of stop words in Natural Language Processing. More specifically, we eliminate DNS queries to common services (e.g., "*pool.ntp.org, time.nist.gov, google.com*, www.google.com"), those that terminate in "*.arpa*", and those that are sent only in the local network (e.g., DNS queries ending in "*.local*") as they may not be observable at the point of packet capture. As a result, dataset1 comprises 10,298 remaining instances from 106 devices, which are in turn from 40 vendors, while dataset 2 consists of 319 instances from 22 devices, and 15 vendors.
- **Training:** We split dataset 1, into a training and testing datasets according to a 80:20 ratio using stratified sampling. Using the training dataset, we then train an attention-based model to predict for each instance the vendor of the IoT device. Embeddings of 256 dimensions are learned for each DNS query; and both encoder and decoder consist of a GRU of 1,024 units. The lengths of the instances vary from 1 to 171,301 DNS queries, with a median of 98 DNS queries. For each instance, we limit the length of the input sequences and consider only the first 100 DNS queries.
- **Extracting:** We apply the *Alexandr* algorithm to the trained attention-based model, and the training dataset to extract rules for each class. We set *window.size* to 1 (Algorithm 1, line 5), and *max_len* to 10 (Algorithm 1, line 15). In other words, although *Alexandr* can learn patterns of different lengths, we set the maximum length of each pattern to 1, and the maximum number of disjunctions to 10. This is because compact rules tend to be easier to interpret, and with each pattern consisting of a single DNS query, a middlebox can process incoming traffic in a stateless manner.
- **Validating:** First, we measure the performance of *Alexandr* on the testing dataset. We compare the *f1-scores* of the initial attention-based model and the extracted rules, and calculate the *fidelity* score of the rules by comparing the percentage of instances from the testing dataset that are classified by the extracted rules as in the same class than the initial attention-based model. Then, to better understand the generalization performance, we compare the performance of the attention-based model, and the extracted rules, on an independent dataset, i.e., dataset 2.

4.3   Results

Table 1 summarizes the results using the train/test split method on dataset 1. The extracted rules decrease the storage requirements, and inference latency, by 4 orders of magnitude, compared to the initial attention-based model. The inference latency represents the time taken to predict the class for every instance

Table 1. Summary of results

	Attention-based model	Extracted rules
Memory	161 MB	2.8 KB
Inference latency (testing)	1,472 s	0.192 s
Average f1-score	0.998	0.995
Fidelity	NA	98.94%

in the testing dataset in a sequential manner. The next paragraphs illustrate the extracted rules, and discuss the results (e.g., f1-score, fidelity) on the testing dataset in more details. Then, we validate the extracted rules on dataset 2.

Vendor	Rule
lifx	v2.broker.lifx.co
netatmo	netcom.netatmo.net OR apicom.netatmo.net
philips	dcp.cpp.philips.com OR bridge.meethue.com
...	...
withings	scalews.withings.net

Fig. 4. Excerpt of extracted rules

Extracted Rules. *Alexandr* extracts a total of 40 rules, one for each class. Figure 4 provides an excerpt of the extracted rules. The first shown rule means that if a device sends a DNS query for *v2.broker.lifx.co*, then the IoT device is from the vendor *lifx*. The second shown rule stipulates that if a device sends a DNS query for *net.com.netatmo.com* or *api.netatmo.com*, then the IoT device is from the vendor *netatmo*. The second rule is of length 2, as it consists of two patterns. Across the 40 extracted rules, 32 of them (80%) have a length less than or equal to 2, and the longest rule has a length of 6.

Precision, Recall, F1-Score. Table 2 reports the precision, recall, f1-score, and their average for the attention-based model, and those for the extracted rules, for the 40 classes on the testing dataset. The attention-based model achieves an average f1-score of 0.998. In comparison, the extracted rules achieve an average f1-score of 0.995. As such, the extracted rules provide interpretability with a minimal loss in classification performance.

Fidelity. The fidelity metric represents how accurately the extracted rules correspond to the initial trained attention-based model. It is typically calculated as the percentage of examples classified by the extracted rules as in the same class than the neural network. However, in our context, an instance might match multiple rules and get classified as belonging to multiple classes. More precisely, out of 2,077 instances in the test dataset, 18 instances are classified as belonging to

Table 2. Comparison of classification with attention-based model, and extracted rules

	Attention-based model			Extracted rules		
	Precision	Recall	f1-score	Precision	Recall	f1-score
amazon	1.000	1.000	1.000	1.000	1.000	1.000
arcsoft	1.000	1.000	1.000	1.000	1.000	1.000
belkin	1.000	1.000	1.000	0.983	0.994	0.989
canary	1.000	1.000	1.000	1.000	1.000	1.000
d-link	1.000	1.000	1.000	1.000	1.000	1.000
ecobee	1.000	1.000	1.000	1.000	1.000	1.000
gracedigital	1.000	1.000	1.000	1.000	1.000	1.000
honeywell	1.000	1.000	1.000	1.000	1.000	1.000
ihome	1.000	1.000	1.000	1.000	1.000	1.000
irobot	1.000	0.952	0.976	1.000	0.952	0.976
lg	1.000	1.000	1.000	1.000	1.000	1.000
lifx	1.000	1.000	1.000	1.000	1.000	1.000
logitech	1.000	1.000	1.000	1.000	1.000	1.000
makerbot	1.000	1.000	1.000	1.000	1.000	1.000
movo	1.000	1.000	1.000	1.000	1.000	1.000
netatmo	1.000	1.000	1.000	1.000	1.000	1.000
netgear	1.000	1.000	1.000	1.000	1.000	1.000
philips	1.000	0.938	0.968	1.000	0.937	0.968
pixstar	1.000	1.000	1.000	1.000	1.000	1.000
roku	1.000	1.000	1.000	1.000	1.000	1.000
skybell	1.000	1.000	1.000	1.000	1.000	1.000
smartlabs	1.000	1.000	1.000	1.000	1.000	1.000
sonos	1.000	1.000	1.000	1.000	0.981	0.990
tp-link	1.000	1.000	1.000	1.000	1.000	1.000
wink	1.000	1.000	1.000	1.000	1.000	1.000
withings	1.000	1.000	1.000	1.000	1.000	1.000
zipta	1.000	1.000	1.000	1.000	1.000	1.000
...	...	...	...	...	...	...
Average	0.999	0.997	0.998	0.997	0.994	0.995

2 classes. The fraction of instances classified by the extracted rules as belonging to more than one class therefore represents only 1% of the instances.

Given that the extracted rules might provide multiple classes for an instance, if one of the provided class is the same as the one predicted by the attention-based model, we count that instance as being classified by the extracted rules as in the same class than the attention-basec model, and we obtain a fidelity score of 98.94% (2,055 out of 2,077 instances) on the testing dataset.

Table 3. Confusion matrix of attention-based model on independent dataset (dataset 2)

		amazon	belkin	ecobee	google	ihome	lifx	netatmo	pixstar	samsung	smartlabs	sonos	tplink	unknown	vera	withings
	amazon	20	0	0	0	0	0	0	0	0	0	0	0	0	0	0
	belkin	0	40	0	0	0	0	0	0	0	0	0	0	0	0	0
	ecobee	0	0	0	0	0	0	0	0	0	0	0	0	0	0	0
	google	0	0	0	6	0	0	0	0	0	0	19	0	0	0	0
	ihome	0	0	0	0	10	0	0	0	0	0	0	0	0	0	0
	lifx	0	0	0	0	0	10	0	0	0	0	0	0	0	10	0
True Labels	netatmo	0	0	0	0	0	0	40	0	0	0	0	0	0	0	0
	pixstar	0	0	0	0	0	0	0	13	0	0	0	0	0	0	0
	samsung	0	0	0	0	0	0	0	0	1	0	0	0	0	23	0
	smartlabs	0	0	0	0	0	0	0	0	0	16	0	0	0	0	0
	sonos	0	0	0	0	0	0	0	0	0	0	0	0	0	0	0
	tplink	0	0	12	0	0	0	0	0	0	0	0	0	0	17	0
	unknown	0	0	0	0	0	0	0	0	0	0	1	0	0	42	0
	vera	0	0	0	0	0	0	0	0	0	0	0	0	0	0	0
	withings	0	0	0	0	0	0	0	0	0	0	0	0	0	15	35

Generalization. In order to better understand the generalization performance, we apply the attention-based model, and the extracted rules on dataset 2. Three vendors from dataset 2 were not present in dataset 1 (*blipcare, hp, invoxia*). We therefore label instances belonging to any of those vendors as *"unknown"*.

Table 3 depicts the confusion matrix of the attention-based model, while Table 4 provides that of the extracted rules. Any instance not matching any of the rules is classified as *"unknown"*. In contrast, the attention-based model classifies all instances in one of the classes on which it was trained. It cannot recognize *"unknown"* devices.

The attention-based model achieves a f1-score of 0.553. The poorer performance can be explained by several factors. First, the training dataset did not encompass all device types for each vendor. For example, although the attention-based model was trained to identify devices from *"google"*, the training dataset did not include any Nest smoke detector, whereas dataset 2 did. As such, instances from the Nest smoke detector were commonly misclassified. The same issue happens with the class *"samsung"*: Dataset 2 contains *"samsung"* camera, whereas the training dataset did not. Second, the attention-based model cannot recognize *"unknown"* devices, but instead classify them as belonging to one of the classes it was trained for (e.g., *"sonos"*, *"vera"*). Because dataset 2

Table 4. Confusion matrix of rules on independent dataset (dataset 2)

	amazon	belkin	google	ihome	lifx	netatmo	pixstar	samsung	smartlabs	tplink	unknown	withings
amazon	20	0	0	0	0	0	0	0	0	0	0	0
belkin	0	40	0	0	0	0	0	0	0	0	0	0
google	0	0	5	0	0	0	0	0	0	0	20	0
ihome	0	0	0	10	0	0	0	0	0	0	0	0
lifx	0	0	0	0	10	0	0	0	0	0	0	0
netatmo	0	0	0	0	0	40	0	0	0	0	0	0
pixstar	0	0	0	0	0	0	13	0	0	0	0	0
samsung	0	0	0	0	0	0	0	1	0	0	23	0
smartlabs	0	0	0	0	0	0	0	0	16	0	0	0
tplink	0	0	0	0	0	0	0	0	0	0	28	0
unknown	0	0	0	0	0	0	0	0	0	0	43	0
withings	0	0	0	0	0	0	0	0	0	0	15	35

(True Labels)

did not in reality present any instance from those classes, their f1-score result in 0.0, decreasing the overall average f1-score.

In contrast, the rules achieve an average f1-score of 0.728 demonstrating that the rules provide improved generalized performance. For few classes (*"google"*, *"samsung"*), the rules resulted in poor performance, similarly to the attention-based model. As explained above, this may be because the testing dataset (dataset 2) including device types from those vendors that were not present in the training dataset. However, for most classes (e.g., *amazon, belkin, ihome, lifx, netatmo, smartlabs, withings*), the rules are performing as well as the attention-based model, and correctly identify unknown devices.

5   Conclusion and Future Work

We introduced a novel efficient algorithm that allows one to extract interpretable rules from a trained attention-based model. Evaluations using a train:test split approach demonstrate that the rules not only reduce the storage requirements, and inference latency, but also provide interpretability and achieve high f1-score, and fidelity, with minimal loss compared to the initial attention-based model. Further evaluation on an independent dataset shows improved generalization performance, and the ability to identify unknown devices.

In the future, we will explore methods to extract not only positive evidence but also negative evidence for the rules. To illustrate the current limitations, a rule *"belkin.com → belkin"* may result in high false positive as not only belkin devices, but users and mobile applications may also send queries to *belkin.com*. Instead, the integration of negative clauses, e.g., *"belkin.com* AND **not** *android.clients.google.com* → *belkin"*, may provide even better performance.

Acknowledgment. The authors would like to thank the anonymous reviewers for their suggestions, and comments.

References

1. Hautala, L.: Why it was so easy to hack the cameras that took down the web. In: CNET Security, October 2016
2. Palmer, D.: 175,000 IoT cameras can be remotely hacked thanks to flaw, says security researcher. In: ZDNet, July 2017
3. Yu, T., Sekar, V., Seshan, S., Agarwal, Y., Xu, C.: Handling a trillion (unfixable) flaws on a billion devices: rethinking network security for the internet-of-things. In: Proceedings of the 14th ACM Workshop on Hot Topics in Networks, HotNets-XIV (2015)
4. Sivanathan, A., et al.: Characterizing and classifying IoT traffic in smart cities and campuses. In: IEEE Infocom Workshop Smart Cities and Urban Computing (2017)
5. Miettinen, M., et al.: Iot sentinel demo: automated device-type identification for security enforcement in iot. In: IEEE ICDCS (2017)
6. Meidan, Y., et al.: Profiliot: a machine learning approach for IoT device identification based on network traffic analysis, April 2017
7. Guo, H., Heidemann, J.: IP-based IoT device detection. In: Proceedings of the 2018 Workshop on IoT Security and Privacy, IoT Samp;P 2018, (New York, NY, USA), pp. 36–42. Association for Computing Machinery (2018)
8. Ortiz, J., Crawford, C., Le, F.: Devicemien: network device behavior modeling for identifying unknown Iot devices. In: Proceedings of the International Conference on Internet of Things Design and Implementation, IoTDI 2019, (New York, NY, USA), pp. 106–117. Association for Computing Machinery (2019)
9. Bremler-Barr, A., Levy, H., Yakhini, Z.: IoT or not: identifying IoT devices in a shorttime scale (2019)
10. Mazhar, M.H., Shafiq, Z.: Characterizing smart home IoT traffic in the wild (2020)
11. Huang, D.Y., Apthorpe, N., Acar, G., Li, F., Feamster, N.: Iot inspector: crowdsourcing labeled network traffic from smart home devices at scale (2019)
12. Paxson, V.: Bro: a system for detecting network intruders in real-time. Comput. Netw. **31**(23–24), 2435–2463 (1999)
13. Roesch, M.: Snort - lightweight intrusion detection for networks. In: Proceedings of the 13th USENIX Conference on System Administration, LISA 1999 (USA), pp. 229–238. USENIX Association (1999)
14. Cleeremans, A., Servan-Schreiber, D., Mcclelland, J.: Finite state automata and simple recurrent networks. Neural Comput. - NECO **1**, 372–381 (1989)
15. Hayashi, Y., Imura, A.: Fuzzy neural expert system with automated extraction of fuzzy if-then rules from a trained neural network. In: Proceedings First International Symposium on Uncertainty Modeling and Analysis (1990)
16. Towell, G.G., Shavlik, J.W.: The extraction of refined rules from knowledge-based neural networks. In: Machine Learning, pp. 71–101 (1993)
17. Fu, L.: Rule generation from neural networks. In: IEEE Transactions on Systems, Man, and Cybernetics (1994)

18. Omlin, C., Giles, C.: Extraction of rules from discrete-time recurrent neural network. Neural Netw. **9**, 41–52 (2001)
19. Murdoch, W.J., Szlam, A.: Automatic rule extraction from long short term memory networks, CoRR, vol. abs/1702.02540 (2017)
20. Ribeiro, M.T., Singh, S., Guestrin, C.: "why should i trust you?": explaining the predictions of any classifier. In: Proceedings of the 22nd ACM SIGKDD International Conference on Knowledge Discovery and Data Mining, KDD 2016, (New York, NY, USA), pp. 1135–1144. Association for Computing Machinery (2016)
21. Bahdanau, D., Cho, K., Bengio, Y.: Neural machine translation by jointly learning to align and translate. arXiv e-prints, p. arXiv:1409.0473, September 2014
22. Xu, K., et al.: Show, attend and tell: neural image caption generation with visual attention. arXiv e-prints, p. arXiv:1502.03044, February 2015
23. Chorowski, J.K., Bahdanau, D., Serdyuk, D., Cho, K., Bengio, Y.: Attention-based models for speech recognition. In: Cortes, C., Lawrence, N.D., Lee, D.D., Sugiyama, M., Garnett, R. (eds.) Advances in Neural Information Processing Systems 28, pp. 577–585. Curran Associates Inc. (2015)
24. Sutskever, I., Vinyals, O., Le, Q.V.: Sequence to sequence learning with neural networks. arXiv e-prints, p. arXiv:1409.3215, September 2014
25. Cho, K., van Merrienboer, B., Gülçehre, Ç., Bougares, F., Schwenk, H., Bengio, Y.: Learning phrase representations using RNN encoder-decoder for statistical machine translation. CoRR, vol. abs/1406.1078 (2014)
26. Rayome, A.D.: Security flaw made 175,000 iot cameras vulnerable to becoming spy cams for hackers. https://www.techrepublic.com/article/security-flaw-made-175000-iot-cameras-vulnerable-to-becoming-spy-cams-for-hackers//. Accessed 01 August 2017

Combining Individual and Joint Networking Behavior for Intelligent IoT Analytics

Jeya Vikranth Jeyakumar[1,4], Ludmila Cherkasova[1(✉)], Saina Lajevardi[2],
Moray Allan[3], Yue Zhao[2], John Fry[2], and Mani Srivastava[4]

[1] Arm Research, San Jose, USA
[2] Arm Inc., San Jose, USA
`ludmila.cherkassova@arm.research`
[3] Arm Inc., Glasgow, UK
[4] University of California, Los Angeles, CA, USA
`vikranth94@g.ucla.edu, mbs@ucla.edu`

Abstract. The IoT vision of a trillion connected devices over the next decade requires reliable end-to-end connectivity and automated device management platforms. While we have seen successful efforts for maintaining small IoT testbeds, there are multiple challenges for the efficient management of large-scale device deployments. With Industrial IoT, incorporating millions of devices, traditional management methods do not scale well. In this work, we address these challenges by designing a set of novel machine learning techniques, which form a foundation of a new tool, *IoTelligent*, for IoT device management, using traffic characteristics obtained at the network level. The design of our tool is driven by the analysis of 1-year long networking data, collected from 350 companies with IoT deployments. The exploratory analysis of this data reveals that IoT environments follow the famous Pareto principle, such as: (i) 10% of the companies in the dataset contribute to 90% of the entire traffic; (ii) 7% of all the companies in the set own 90% of all the devices. We designed and evaluated CNN, LSTM, and Convolutional LSTM models for demand forecasting, with a conclusion of the Convolutional LSTM model being the best. However, maintaining and updating individual company models is expensive. In this work, we design a novel, scalable approach, where a general demand forecasting model is built using the combined data of all the companies with a normalization factor. Moreover, we introduce a novel technique for device management, based on autoencoders. They automatically extract relevant device features to identify device groups with similar behavior to flag anomalous devices.

Keywords: Forecasting · Deep learning · Device management

1 Introduction

The high-tech industry expects a trillion new IoT devices will be produced between now and 2035 [3,6,23]. These devices could range from simple sensors

© Springer Nature Switzerland AG 2020
W. Song et al. (Eds.): ICIOT 2020, LNCS 12405, pp. 45–62, 2020.
https://doi.org/10.1007/978-3-030-59615-6_4

in everyday objects to complex devices, defined by the industrial and manufacturing processes. The Internet of Things ecosystem should include the necessary components that enable businesses, governments, and consumers to seamlessly connect to their IoT devices. This vision requires reliable end-to-end connectivity and device management platform, which makes it easier for device owners to access their IoT data and exploiting the opportunity to derive real business value from this data. The benefits of leveraging this data are greater business efficiencies, faster time to market, cost savings, and new revenue streams. Embracing these benefits ultimately comes down to ensuring the data is secure and readily accessible for meaningful insights.

The Arm Mbed IoT Device Management Platform [4] addresses these requirements by enabling organizations to securely develop, provision and manage connected devices at scale and by enabling the connectivity management [5] of every device regardless of its location or network type. The designed platform supports the physical connectivity across all major wireless protocols (such as cellular, LoRa, Satellite, etc.) that can be managed through a single user interface. Seamlessly connecting all IoT devices is important in ensuring their data is accessible at the appropriate time and cost across any use case. While we could see successful examples of deploying and maintaining small IoT testbeds, *there are multiple challenges in designing an efficient management platform for large-scale device deployments.* The operators of IoT environments may not be fully aware of their IoT assets, let alone whether each IoT device is functioning and connected properly, and whether enough networking resources and bandwidth allocated to support the performance objectives of their IoT networks. With the IoT devices being projected to scale to billions, the traditional (customized or manual) methods of device and IoT networks management do not scale to meet the required performance objectives.

In this work, we aim to address these challenges by designing a set of novel machine learning techniques, which form a foundation of a new tool, *IoTelligent*, for IoT networks and device management, using traffic characteristics obtained at the network level. One of the main objectives of *IoTelligent* is to build effective demand forecasting methods for owners of IoT ecosystems to manage trends, predict performance, and detect failures. The insights and prediction results of the tool will be of interest to the operators of IoT environments.

For designing the tool and appropriate techniques, we utilize the unique set of real (anonymized) data, which were provided to us by our business partners. This dataset represents 1-year of networking data collected from 350 companies with IoT deployments, utilizing the Arm Mbed IoT Device Management Platform. The exploratory analysis of the underlying dataset reveals a set of interesting insights into the nature of such IoT deployments. It shows that the IoT environments exhibit properties similar to the earlier studied web and media sites [8, 9, 20, 28] and could be described by famous Pareto principle [1], when the data distribution follows the power law [2]. The Pareto principle (also known as the 80/20 rule or the "law of the vital few") states that for many events or data distributions roughly 80% of the effects come from 20% of the causes.

For example, in the earlier web sites, 20% of the web pages were responsible for 80% of all the users accesses [8]. The later, popular web sites follow a slightly different proportion rule: they often are described by 90/10 or 90/5 distributions, i.e., 90% of all the user accesses are targeting a small subset of popular web pages, which represent 5% or 10% of the entire web pages set.

The interesting findings from the studied IoT networking dataset can be summarized as follows:

- 10% of the companies in the dataset contribute to 90% of the entire traffic;
- 7% of all the companies in the dataset own 90% of all the devices.

IoTelligent tool applies machine learning techniques to forecast the companies' traffic demands over time, visualize traffic trends, identify and cluster devices, detect device anomalies and failures. We designed and evaluated CNN, LSTM, and Convolutional LSTM models for demand forecasting, with a conclusion of the Convolutional LSTM model being the best. To avoid maintaining and upgrading tens (or hundreds) of models (a different model per company), we designed and implemented a novel, scalable approach, where a global demand forecasting model is built using the combined data of all the companies. The accuracy of the designed approach is further improved by normalizing the "contribution" of individual company data in the combined global dataset. To solve the scalability issues with managing the millions of devices, we designed and evaluated a novel technique based on: (i) autoencoders, which extract the relevant features automatically from the network traffic stream; (ii) DBSCAN clustering to identify the group of devices that exhibit similar behavior, to flag anomalous devices. The designed management tool paves the way the industry can monitor their IoT assets for presence, functionality, and behavior at scale without the need to develop device-specific models.

2   Dataset and the Exploratory Data Analysis

The network traffic data was collected from more than 350 companies for a total duration of one year. The traffic data is binned using 15 min time window, used for billing purposes.

- Unix timestamp;
- Anonymous company ids;
- Anonymous device ids per company;
- The direction of the network traffic (to and from the device);
- Number of bytes transmitted in the 15 min interval;
- Number of packets transmitted in the 15 min interval.

Preliminary analysis was done to find the most impactful and well-established companies. We found that the companies' data that represent two essential metrics, such as the networking traffic amount and number of deployed IoT devices, both follow the Pareto law. The main findings from the studied IoT networking dataset can be summarized as follows:

– 10% of the companies in the dataset contribute to 90% of the entire traffic;
– 7% of all the companies in the dataset own 90% of all the devices.

Figure 1 shows on the left side the logscale graph of CDF (Cumulative Distribution Function) of the traffic (where one can see that 10% of the companies in the dataset contribute to 90% of the entire traffic) and the CDF of the devices per company distribution (where one can see that 7% of all the companies in the dataset own 90% of all the devices). Also, it is quite interesting to note how significant and sizable the contributions are of the first 5–10 companies on those graphs: both for the networking traffic volume and the number of overall devices.

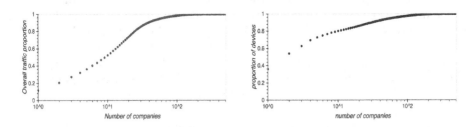

Fig. 1. (left) CDF of networking traffic; (right) CDF of devices.

Another interesting observation was that companies with highest number of devices did not correspond to companies with maximum amount of traffic, and vice versa, the high volume traffic companies did not have a lot of devices (Fig. 2). This makes sense, for example, a difference in the outputs of hundreds of simple sensors and a single recording camera. Among some other insights into special properties of many IoT environments (at the networking level) we observe the pronounced diurnal and weekly patterns, and changes in the traffic patterns around some seasonal events and holidays. It could be explained by the fact that many IoT environments are related to human and business activities.

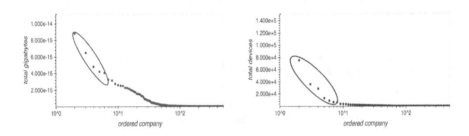

Fig. 2. (left) Networking traffic per company; (right) Number of devices per company.

3   Demand Forecasting

The *demand forecasting problem* is formulated in the following way. Given a recent month's traffic pattern for a company, what is the expected traffic for this company a week ahead? This problem requires that a predictive model forecasts the total number of bytes for each hour over the next seven days. Technically, this framing of the problem is referred to as a multi-step time series forecasting problem, given the multiple forecast steps. Choosing the right time granularity for (i) making the prediction and (ii) data used in the model, is another important decision for this type of a problem.

We found that a reasonable trade-off would be to use 1 h time granularity. This eliminates the small noises in traffic and also ensures that we have a sufficient data to train our models on.

3.1   Modeling Approach

Based on our exploratory data analysis, we select 33 companies with largest traffic and 5 companies with largest number of devices. These companies are responsible for 90% of the networking traffic volume and 90% of IoT devices. Therefore, by designing and evaluating the modeling approach for these companies, we could efficiently cover the demand forecasting for 90% of the traffic volume and assessing the monitoring solution for 90% of devices.

The specific goal is to predict the company traffic for a next week given the previous three weeks of traffic data in an hourly time granularity. We use **a deep learning** based approach for demand forecasting, because deep learning methods are robust to noise, highly scalable, and generalizable. We have considered three different deep learning architectures for demand forecasting: CNN, LSTM, and Convolutional LSTM in order to compare their outcome and accuracy.

Convolutional Neural Network (CNN) [21]: It is a biologically inspired variant of a fully connected layer, which is designed to use minimal amounts of preprocessing. CNNs are made of Convolutional layers that exploit spatially-local correlation by enforcing a local connectivity pattern between neurons of adjacent layers. The main operations in Convolution layers are Convolution, Activation (ReLU), Batch normalization, and Pooling or Sub-Sampling. The CNN architecture, used in our experiments, has 4 main layers. The first three layers are one-dimensional convolutional layers, each with 64 filters and relu activation function, that operate over the 1D traffic sequence. Each convolutional layer is followed by a max-pooling layer of size 2, whose job is to distill the output of the convolutional layer to the most salient elements. A flatten layer is used after the convolutional layers to reduce the feature maps to a single one-dimensional vector. The final layer is a dense fully connected layer with 168 neurons (24 h × 7 d) with linear activation and that produces the forecast by interpreting the features extracted by the convolutional part of the model.

Long Short Term Memory (LSTM) [13,15]: It is a type of Recurrent Neural Network (RNN), which takes current inputs and remembers what it has perceived previously in time. An LSTM layer has a chain-like structure of repeating units and each unit is composed of a cell, an input gate, an output gate, and a forget gate, working together. It is well-suited to classify, process, and predict time series with time lags of unknown size and duration between important events. Because LSTMs can remember values over arbitrary intervals, they usually have an advantage over alternative RNNs, Hidden Markov models, and other sequence learning methods in numerous applications. The model architecture, used in our experiments, consists of two stacked LSTM layers, each with 32 LSTM cells, followed by a dense layer with 168 neurons to generate the forecast.

Convolutional LSTM [30]: Convolutional LSTM is a hybrid deep learning architecture that consists of both convolutional and LSTM layers. The first two layers are the one-dimensional Convolutional layers that help in capturing the high-level features from the input sequence of traffic data. Each convolutional layer is followed by a max-pooling layer to reduce the sequence length. They are followed by two LSTM layers, that help in tracking the temporal information from the sequential features, captured by the convolutional layers. The final layer is a dense fully connected layer, that gives a forecasting output.

We use batch-normalization and dropout layers in all our models. To evaluate the prediction accuracy of the designed models, we compare the predicted value X_n^{pred} with the true, measured value X_n using following error metrics:

Mean Absolute Error (MAE):

$$MAE = \frac{1}{N} \Sigma_{n=1}^{N} |X_n - X_n^{pred}|$$

Mean Squared Error (MSE):

$$MSE = \frac{1}{N} \Sigma_{n=1}^{N} (X_n - X_n^{pred})^2$$

3.2   Individual Model per Company

This is the naive approach where each company has it's own demand forecasting model, that is, the model for each company is trained by using only the data from that particular company as shown in Fig. 3 (a).

So, for each company, we trained three models with the architectures described above (i.e., CNN, LSTM, and Convolutional LSTM). Figure 4 (a) presents the detailed parameters of the designed Convolutional LSTM, while Fig. 4 (b) reflects the relative performance of three different architectures (with the MAE error metrics). We found that for both error metrics the Convolutional LSTM model performs better than the other two architectures. When comparing architectures' accuracy by using MAE and MSE, we can see that Convolutional LSTM outperforms CNN by **16%** and **23%** respectively, and outperforms LSTM by **43%** and **36%** respectively. Therefore, only **Convolutional LSTM** architecture is considered for the rest of the paper. Finally, Fig. 5 shows an example of company A (in the studied dataset): its measured networking traffic over time and the forecasting results with the Convolutional LSTM model.

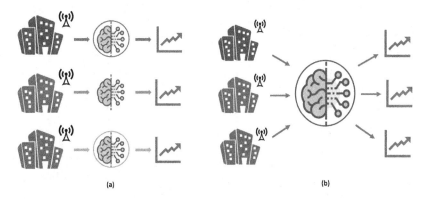

Fig. 3. (a) Each company has its own prediction model, (b) Using one model for all the companies, trained on the combined dataset.

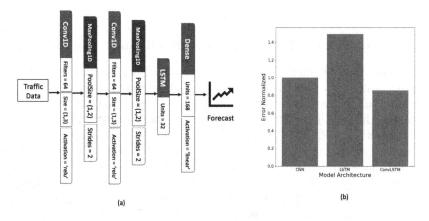

Fig. 4. (a) Model Architecture of Convolutional LSTM Model; (b) Comparing performance of the three architectures: Convolutional LSTM achieves best performance.

Building an individual model per each company has a benefit that this approach is simple to implement. But this method has significant drawbacks. First, the model easily overfits on the training data since it's trained only using limited data from a particular company resulting in poor forecasting performance. Secondly, it is not scalable as the number of required forecasting models is directly proportional to the number of companies. The service provider has to deal with the models' maintenance, their upgrades, and retraining (with new data) over time. Therefore, in the next Sect. 3.3, we aim to explore a different approach, which enables a service provider to use all the collected, global data for building a single (global) model, while using it for individual company demand forecasting. Only Convolutional LSTM architecture is considered in the remaining of the paper (since as shown, it supports the best performance).

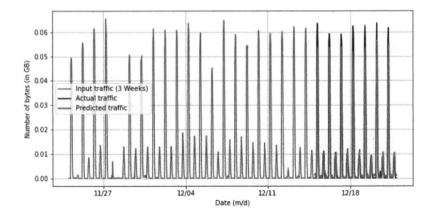

Fig. 5. Company a: the 4th week demand forecast based on data from the previous 3 weeks.

3.3   One Model for All Companies - Without Normalization

In this approach, we train a single Convolutional LSTM model for demand forecasting by using data from all the companies. The networking traffic data from all the companies were combined. The data from January to October were used for training the model, and the data from November and December were used as the test set.

This method is highly scalable since it trains and utilizes a single model for demand forecasting of all companies. While this approach is very attractive and logical, it did not always produce good forecasting results. Fig. 6 shows the forecasting made by this global model for Company A (with this company we are already familiar from Fig. 5). As we can see in Fig. 6, the model fails to capture a well-established traffic pattern.

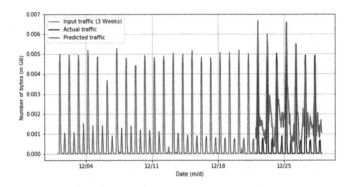

Fig. 6. Demand forecasting using the Global model trained on data without normalization.

One of the explanations of the observed issue is that this company's traffic constitutes a very small fraction compared to the other companies in the combined dataset. So, the globally trained model has "learned" the traffic patterns of larger companies in the set, while has "downplayed" the traffic patterns of smaller companies. The reason the model fails to capture a well-established traffic pattern for companies with less traffic is because, the traffic prediction loss in terms of absolute value is still small. However, it is not a desirable outcome as we would like our model to capture the traffic pattern even for companies with low traffic.

3.4   One Model for All Companies - with Normalization

This method aims to address the issues of the previous two approaches. In this method, the data from each company is normalized, that is, all the data subsets are scaled so that they lie within the same range. We use the min-max scaling approach to normalize the data subsets so that the values of the data for all companies lie between 0 and 1. Equation 1 shows the formula used for min-max scaling, where 'i' refers to the 'i'th company.

$$X^i_{norm} = \frac{X^i - X^i_{min}}{X^i_{max} - X^i_{min}} \tag{1}$$

Then a single deep learning model for forecasting is trained using the normalized data of all companies. The predicted demand (forecast) is then re-scaled using Eq. 2 to the original scale.

$$X^i = X^i_{norm} * (X^i_{max} - X^i_{min}) + X^i_{min} \tag{2}$$

This method of training the global model gives equal importance to the data from all companies and treats them fairly. The designed model does not over-fit and is generalizable since it is trained on the data from multiple companies. Figure 7 graphically reflects the process of the global model creation with normalized data from different companies.

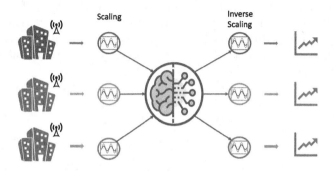

Fig. 7. One global prediction model is trained by using the normalized data from all the companies.

Figure 8 shows that the designed forecasting model can capture well the patterns of companies with low traffic volume (such as Company A).

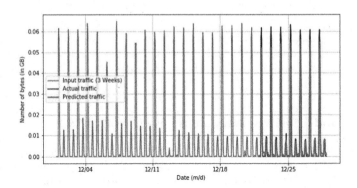

Fig. 8. Demand forecasting using the Global model trained on data with normalization.

4   Introducing Uncertainty to Forecasting Models

In the previous section, we designed a single global model with normalization, that can be used to forecast for multiple companies. But demand forecasting is a field, where an element of uncertainty exists in all the predictions, and therefore, representing model uncertainty is of crucial importance. The standard deep learning tools for forecasting do not capture model uncertainty. Gal et al. [12] propose a simple approach to quantify the neural network uncertainty, which shows that the use of dropout in neural networks can be interpreted as a Bayesian approximation of a Gaussian process - a well known probabilistic model. Dropout is used in many models in deep learning as a way to avoid over-fitting, and it acts as a regularizer. However, by leaving it "on" during the prediction, we end up with the equivalent of an ensemble of subnetworks, within our single larger network, that have slightly different views of the data. If we create a set of T predictions from our model, we can use the mean and variance of these predictions to estimate the prediction set uncertainty. Figure 9 shows the forecast with uncertainty for Company A, using the global model with normalization.

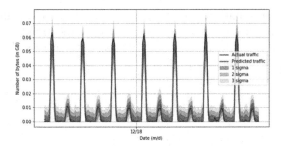

Fig. 9. Demand forecasting with uncertainty for Global model trained on data with normalization.

To evaluate the quality of forecast based on uncertainty, we introduce Prediction Interval Coverage Probability (PICP) metric.

4.1 Prediction Interval Coverage Probability (PICP)

PICP tells us the percentage of time an interval contains the actual value of the prediction. Equations 3–5 show the calculation of PICP metric, where l is the lower bound, u is the upper bound, x_i is the value at timestep i, $\hat{y}$ is the mean of the predicted distribution, z is the number of standard deviations from the Gaussian distribution, (e.g., 1.96 for a 95% interval), and σ is the standard deviation of the predicted distribution.

$$l(x_i) = \hat{y}_i - z * \sigma_i \tag{3}$$

$$u(x_i) = \hat{y}_i + z * \sigma_i \tag{4}$$

$$PICP_{l(x),u(x)} = \frac{1}{N} \sum_{i=1}^{N} h_i, \quad where \; h_i = \begin{cases} 1, & if \; l(x_i) \le y_i \le u(x_i) \\ 0, & otherwise \end{cases} \tag{5}$$

4.2 Evaluating Forecast with Uncertainty

We evaluate the overall performance of our global forecast model, introduced in Sect. 3.4, based on the PICP metric described above. The forecasting is done 100 times for each company with a dropout probability of 0.2, and then the mean and standard deviations are obtained for each company. Figure 10 shows the global model's forecast for the third week of December for two different companies: Company B and Company C. As we can see from the plot, the model captures the traffic pattern, but still, the predicted values show some deviations from the actual values. This results in some errors when using the traditional error metrics discussed in Sect. 3.1, though the model is performing very well. Therefore, introducing uncertainty helps the model to generate a reasonable forecast distribution. Figure 11 shows the forecast with uncertainty, where the different shades of blue indicate the uncertainty interval, obtained for different values of uncertainty multipliers. As we can see from the plot, the single global forecasting model can capture well the general traffic trends across multiple companies. Figure 12 shows the mean PICP calculated across all the companies for the different uncertainty multipliers.

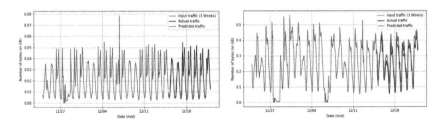

Fig. 10. Demand forecasting with Global model trained on data with normalization.

We find that on an average 50% of the forecast values lie within the predicted interval with one standard deviation, 74% for two standard deviations and 85% for three standard deviations. The forecast samples which lied outside the predicted interval were mostly due to the fact that the months of November and December had lots of holidays and hence those days did not follow the captured traffic pattern.

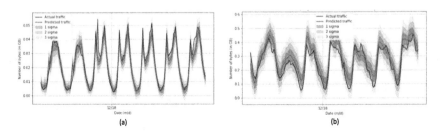

Fig. 11. Demand forecasting with uncertainty using the Global model trained on data with normalization.

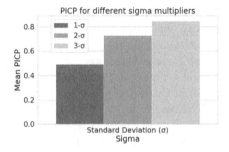

Fig. 12. Mean PICP for different values of sigma multiplier.

5   Device Monitoring and Diagnostics

Once an Internet of Things (IoT) ecosystem is installed, it does not follow a "fire and forget" scenario. There will be unforeseen operational issues, some devices will fail and/or would need to be either repaired or replaced. Each time this happens, the company is on the mission to minimize the downtime and ensure that its devices function properly to protect their revenue stream. However, to address the issues of failed or misbehaving devices, we need to identify such devices in the first place. Therefore, the ability to monitor the device's health and being able to detect, when something is amiss, such as higher-than-normal network traffic or "unusual" device behavior, it is essential to proactively identify and diagnose potential bugs/issues. Again, large-scale IoT deployments is a critical and challenging issue. When there are thousands of devices in the IoT ecosystem, it becomes extremely difficult to efficiently manage these devices as it is practically impossible to monitor each device individually. So, we need an efficient way to analyze the observed device behaviors and identify devices that show an anomalous ("out of usual norm") behavior.

Anomalous or failed devices can be categorized into two types:

1. The devices that behave significantly different from the other devices;
2. The devices whose observed behavior suddenly changes from its "normal" behavior over time.

The following Section describes the designed technique to accomplish the device monitoring and diagnostic via device categorization over time.

5.1   Cluster Devices Based on Their Traffic Patterns and Identify Anomalous Devices

When there are thousands of devices in a given IOT Ecosytem, there usually exist multiple devices of the same type or having similar behavior. We identify these groups of devices in an unsupervised manner based on their network traffic pattern over a given month. Figure 13 shows an overview of the proposed method and its steps to obtain the groups of "similar" devices:

– The monthly network traffic from the thousands of IoT devices are passed through an autoencoder to extract features in the latent space in an unsupervised manner.
– Then we use a density-based clustering algorithm, DBSCAN, on the latent space to identify the groups of similar devices. The objective is to learn what normal data points looks like and then use that to detect abnormal instances. Any instance that has a low affinity to all the clusters is likely to be an anomaly.

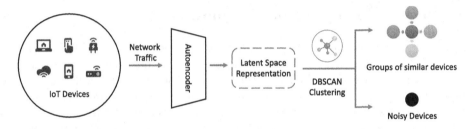

Fig. 13. The pipeline for clustering similar groups of devices and detecting noisy devices based on their traffic patterns.

Autoencoder [22]: It is a neural network capable of learning dense representations of the input data, called latent space representations, in an unsupervised manner. The latent space has low dimensions which helps in visualization and dimensionality reduction. An autoencoder has two parts: an encoder network that encodes the input values x, using an encoder function f, and, a decoder network that decodes the encoded values f(x), using a decoder function g, to create output values identical to the input values. Autoencoder's objective is to minimize reconstruction error between the input and output. This helps autoencoders to capture the important features and patterns present in the data in a low dimensional space. When a representation allows a good reconstruction of its input, then it has retained much of the information present in the input. In our experiment, an autoencoder is trained using the monthly traffic data from the IoT devices which captures the important features or the encoding of the devices in the latent space.

Architecture of the Autoencoder: We use a stacked autoencoder in our experiment with two fully connected hidden layers each in the encoder and the decoder. The central bottle neck layer was a fully connected layer with just three neurons which helps in reducing the dimensions. We used mean squared error as the reconstruction loss function.

DBSCAN [11]: (Density-Based Spatial Clustering of Applications with Noise), is a density-based clustering algorithm that captures the insight that clusters are dense groups of points. If a particular point belongs to a cluster, it should be near to lots of other points in that cluster. The algorithm works in the following order: First, we choose two parameters, a positive number, epsilon and a natural number, minPoints. We then begin by picking an arbitrary point in our dataset. If there are more than minPoints points within a distance of epsilon from that point, (including the original point itself), we consider all of them to be part of a "cluster". We then expand that cluster by checking all of the new points and seeing if they too have more than minPoints points within a distance of epsilon, growing the cluster recursively if so. Eventually, we run out of points to add to the cluster. We then pick a new arbitrary point and repeat the process. Now, it's entirely possible that a point we pick has fewer than minPoints points in its epsilon ball, and is also not a part of any other cluster. If that is the case, it's considered a "noise point" not belonging to any cluster and we mark that as an anomaly.

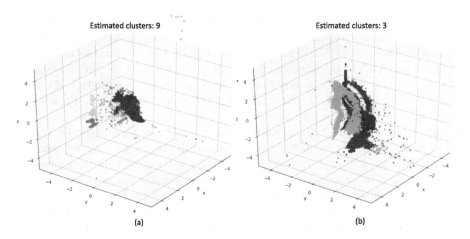

Fig. 14. Number of clusters per company when visualized in the latent space. The black points represent the anomalous devices

Figure 14 shows the latent space and the clusters obtained for Company A(left) and Company B(right). Companies A and B had more than 30000 devices each, installed in their IoT ecosystems and they had three and nine unique types of devices respectively. Based on the devices' traffic patterns observed over the period of a month, the autoencoder mapped the devices of the same type close to each other while the devices of different types were mapped far apart from each other in the latent space. When DBSCAN clustering was applied in the latent space, we observed that the number of distinct clusters formed was exactly the same as the corresponding number of device types per company. The devices which didn't fall in these well formed clusters because of their different traffic patterns were marked as anomalies and are represented by the black points.

6 Related Work

Demand forecasting has been broadly studied due to the problem importance and its significance for utility companies. *Statistical methods* use historical data to make the forecast as a function of most significant variables. The detailed survey on regression analysis for the prediction of residential energy consumption is offered in [7] where the authors believe that among statistical models, linear regression analysis has shown promising results because of satisfactory accuracy and simpler implementation compared to other methods. In many cases, the choice of the framework and the modeling efforts are driven by the specifics of the problem formulation.

While different studies have shown that demand forecasting depends on multiple factors and hence can be used in multivariate modeling, the univariate methods like ARMA and ARIMA [24,25] might be sufficient for short term forecast. *Machine learning* (ML) and *artificial intelligence* (AI) methods based on neural networks [17,27], support vector machines SVM) [10], and fuzzy logic [26] were applied to capture complex non-linear relationships between inputs and outputs. When comparing ARIMA, traditional machine learning, and artificial neural networks (ANN) modeling, some recent articles provide contradictory results. In [16], ARIMA achieves better results than ANN, while the study [14] claims that ANNs perform slightly better than ARIMA methods. In our work, we construct a deep-learning based Convolutional LSTM forecasting model (a hybrid model with both Convolutional and LSTM layers). The Convolutional LSTM model works well on time series data as shown in [18,19] for activity recognition and in [30] for forecasting rainfall. They are good in long term demand prediction, and indeed, automatically captures non-linear patterns.

In general, the quality and the prediction power of the models designed by using ML and AI methods critically depend on the quality and quantity of historical data. To create a good forecasting model, several approaches have been developed in the literature. One such approach is an ensemble of multiple forecasting methods applied on the same time series data and a weighted average of their forecasts is used as a final result [29]. In our work, we pursue a different approach by making use of the normalized data from multiple companies and train a single global model to make traffic predictions. This makes our method highly scalable.

7   Conclusion

In our work, we proposed *IoTelligent*, a tool that applies machine learning techniques to forecast the companies' traffic demands over time, visualize traffic trends, identify and cluster devices, detect device anomalies and failures. We showed that among the different neural network architectures, Convolutional LSTM model performed the best for demand forecasting. In order to avoid maintaining and upgrading tens (or hundreds) of models (a different model per company), we designed and implemented a novel, scalable approach, where a global demand forecasting model is built using the combined data of all the companies. This method was improved by normalizing the "contribution" of individual company data in the combined global dataset. We also introduced uncertainty intervals to the forecasts to provide better information to the users. To solve the scalability issues with managing the millions of devices, we designed and evaluated a novel technique based on: (i) autoencoders, which extract the relevant features automatically from the network traffic stream; (ii) DBSCAN clustering to identify the group of devices that exhibit similar behavior, in order to flag anomalous devices. The designed management tool paves the way the industry can monitor their IoT assets for presence, functionality, and behavior at scale without the need to develop device specific models.

Acknowledgments. This work was completed during J.V. Jeyakumar's summer internship in 2019 at Arm Research. J.V. Jeyakumar and Prof. M. Srivastava are partially supported by the CONIX Research Center, one of six centers in JUMP, a Semiconductor Research Corporation (SRC) program sponsored by DARPA.

References

1. Pareto principle — Wikipedia, the free encyclopedia. https://en.wikipedia.org/wiki/Pareto_principle. Accessed 25 Apr 2020
2. Power law — Wikipedia, the free encyclopedia. https://en.wikipedia.org/wiki/Power_law. Accessed 25 Apr 2020
3. Softbank ceo masayoshi son sees a future with 1 trillion internet of things devices (2016). https://venturebeat.com/2016/10/25/softbank-ceo-masayoshi-son-sees-a-future-with-1-trillion-internet-of-things-devices/
4. Arm pelion device management - deploying and managing IoT devices at large scale (2020). https://www.pelion.com/iot-device-management/
5. Arm pelion IoT platform: Connectivity management (2020). https://www.arm.com/products/iot/pelion-iot-platform/connectivity-management
6. One trillion new IoT devices will be produced by 2035 (2020). https://learn.arm.com/route-to-trillion-devices.html
7. Amin, P., Cherkasova, L., Aitken, R., Kache, V.: Analysis and demand forecasting of residential energy consumption at multiple time scales. In: 2019 IFIP/IEEE Symposium on Integrated Network and Service Management (IM), pp. 494–499. IEEE (2019)
8. Arlitt, M.F., Williamson, C.L.: Web server workload characterization: the search for invariants. ACM SIGMETRICS Perform. Eval. Rev. **24**(1), 126–137 (1996)
9. Cherkasova, L., Gupta, M.: Analysis of enterprise media server workloads: access patterns, locality, content evolution, and rates of change. IEEE/ACM Trans. Netw. **12**(5), 781–794 (2004)
10. China, P.R.: Electricity consumption prediction based on data mining techniques with particle swarm optimization. Int. J. Database Theory Appl. **6**(5), 153–164 (2013)
11. Ester, M., Kriegel, H.P., Sander, J., Xu, X., et al.: A density-based algorithm for discovering clusters in large spatial databases with noise. In: Kdd, vol. 96, pp. 226–231 (1996)
12. Gal, Y., Ghahramani, Z.: Dropout as a Bayesian approximation: representing model uncertainty in deep learning. In: International Conference on Machine Learning, pp. 1050–1059 (2016)
13. Gers, F.A., Schmidhuber, J., Cummins, F.: Learning to forget: continual prediction with LSTM (1999)
14. Gerwig, C.: Short term load forecasting for residential buildings—an extensive literature review. In: Neves-Silva, R., Jain, L.C., Howlett, R.J. (eds.) Intelligent Decision Technologies. SIST, vol. 39, pp. 181–193. Springer, Cham (2015). https://doi.org/10.1007/978-3-319-19857-6_17
15. Hochreiter, S., Schmidhuber, J.: Long short-term memory. Neural Comput. **9**(8), 1735–1780 (1997)
16. Høverstad, B.A., Tidemann, A., Langseth, H.: Effects of data cleansing on load prediction algorithms. In: 2013 IEEE Computational Intelligence Applications in Smart Grid (CIASG), pp. 93–100. IEEE (2013)

17. Hu, Y.C.: Electricity consumption prediction using a neural-network-based grey forecasting approach. J. Oper. Res. Soc. **68**(10), 1259–1264 (2017)
18. Jeyakumar, J.V., Lai, L., Suda, N., Srivastava, M.: SenseHAR: a robust virtual activity sensor for smartphones and wearables. In: Proceedings of the 17th Conference on Embedded Networked Sensor Systems, pp. 15–28 (2019)
19. Jeyakumar, J.V., Lee, E.S., Xia, Z., Sandha, S.S., Tausik, N., Srivastava, M.: Deep convolutional bidirectional LSTM based transportation mode recognition. In: Proceedings of the 2018 ACM International Joint Conference and 2018 International Symposium on Pervasive and Ubiquitous Computing and Wearable Computers, pp. 1606–1615 (2018)
20. Kim, W.: Characterizing the scalability of a large web-based shopping system. ACM Trans. Internet Technol. (TOIT) **1**(1), 44–69 (2001)
21. Krizhevsky, A., Sutskever, I., Hinton, G.E.: ImageNet classification with deep convolutional neural networks. In: Advances in Neural Information Processing Systems, pp. 1097–1105 (2012)
22. Masci, J., Meier, U., Cireşan, D., Schmidhuber, J.: Stacked convolutional auto-encoders for hierarchical feature extraction. In: Honkela, T., Duch, W., Girolami, M., Kaski, S. (eds.) ICANN 2011. LNCS, vol. 6791, pp. 52–59. Springer, Heidelberg (2011). https://doi.org/10.1007/978-3-642-21735-7_7
23. Mukkamala, H.: Arm expands IoT connectivity and device management capabilities with stream technologies acquisition (2018). https://www.arm.com/company/news/2018/06/arm-expands-iot-connectivity-and-device-management-capabilities-with-stream-technologies-acquisition
24. Nichiforov, C., Stamatescu, I., Făgărăşan, I., Stamatescu, G.: Energy consumption forecasting using ARIMA and neural network models. In: 2017 5th International Symposium on Electrical and Electronics Engineering (ISEEE), pp. 1–4. IEEE (2017)
25. Nowicka-Zagrajek, J., Weron, R.: Modeling electricity loads in California: ARMA models with hyperbolic noise. Signal Process. **82**(12), 1903–1915 (2002)
26. Song, K.B., Baek, Y.S., Hong, D.H., Jang, G.: Short-term load forecasting for the holidays using fuzzy linear regression method. IEEE Trans. Power Syst. **20**(1), 96–101 (2005)
27. Sulaiman, S., Jeyanthy, P.A., Devaraj, D.: Artificial neural network based day ahead load forecasting using smart meter data. In: 2016 Biennial International Conference on Power and Energy Systems: Towards Sustainable Energy (PESTSE), pp. 1–6. IEEE (2016)
28. Tang, W., Fu, Y., Cherkasova, L., Vahdat, A.: Modeling and generating realistic streaming media server workloads. Comput. Netw. **51**(1), 336–356 (2007)
29. Wang, X., Smith-Miles, K., Hyndman, R.: Rule induction for forecasting method selection: meta-learning the characteristics of univariate time series. Neurocomputing **72**(10–12), 2581–2594 (2009)
30. Xingjian, S., Chen, Z., Wang, H., Yeung, D.Y., Wong, W.K., Woo, W.c.: Convolutional LSTM network: a machine learning approach for precipitation nowcasting. In: Advances in Neural Information Processing Systems, pp. 802–810 (2015)

RelIoT: Reliability Simulator for IoT Networks

Kazim Ergun[1]([envelope]), Xiaofan Yu[1], Nitish Nagesh[2], Ludmila Cherkasova[3], Pietro Mercati[4], and Tajana Rosing[1]

[1] University of California San Diego, La Jolla, USA
{kergun,x1yu,tajana}@ucsd.edu
[2] Technical University of Munich, Munich, Germany
nitish.nagesh@tum.de
[3] Arm Research, San Jose, USA
lucy.cherkasova@arm.com
[4] Intel Corporation, Hillsboro, USA
{pietro.mercati,raid.ayoub}@intel.com

Abstract. The next era of the Internet of Things (IoT) calls for a large-scale deployment of edge devices to meet the growing demands of applications such as smart cities, smart grids, and environmental monitoring. From low-power sensors to multi-core platforms, IoT devices are prone to failures due to the reliability degradation of electronic circuits, batteries, and other components. As the network of heterogeneous devices expands, maintenance costs due to system failures become unmanageable, making reliability a major concern. Prior work has shown the importance of automated reliability management for meeting lifetime goals for individual devices. However, state-of-the-art network simulators do not provide reliability modeling capabilities for IoT networks.

In this paper, we present an integrated reliability framework for IoT networks based on the ns-3 simulator. The lack of such tools restrained researchers from doing reliability-oriented analysis, exploration, and predictions early in the design cycle. Our contribution facilitates this, which can lead to the design of new network reliability management strategies. The proposed framework, besides reliability, incorporates three other interrelated models - power, performance, and temperature - which are required to model reliability. We validate our framework on a mesh network with ten heterogeneous devices, of three different types. We demonstrate that the models accurately capture the power, temperature, and reliability dynamics of real networks. We finally simulate and analyze two examples of energy-optimized and reliability-optimized network configurations to show how the framework offers an opportunity for researchers to explore trade-offs between energy and reliability in IoT networks.

1 Introduction

The Internet of Things (IoT) is a growing network of heterogeneous devices, combining residential, commercial, and industrial domains. Devices range from low-power sensors with limited computational capabilities to multi-core platforms

© Springer Nature Switzerland AG 2020
W. Song et al. (Eds.): ICIOT 2020, LNCS 12405, pp. 63–81, 2020.
https://doi.org/10.1007/978-3-030-59615-6_5

on the high-end. From small scale (e.g., smart homes) to large scale (e.g., smart cities) applications, the IoT provides infrastructure and services for enhancing the quality of life and use of resources. By 2025, the IoT is expected to connect 41 billion devices [14].

The unprecedented scale and heterogeneity of the IoT pose major research challenges that have not been faced before. While ongoing research efforts aim at optimizing power efficiency and performance [29,30], an aspect that has often been neglected is the reliability of the devices in the network. The common property for these devices is that they age, degrade and eventually require maintenance in the form of repair, component replacement, or complete device replacement. Since an enormous number of heterogeneous devices are interconnected in IoT networks, the maintenance costs will increase accordingly. Cisco recently anticipated that for 100K devices that operate IoT smart homes, around $6.7M/year will be spent for administration and technical diagnosis related to system failures, comprising between 30% to 70% of total costs [7]. Without a proper reliability management strategy, IoT solutions are strongly limited as it becomes infeasible to maintain increasingly large networks.

Exploring reliability management strategies requires a convenient tool for reliability evaluation. In respect of this, simulators are widely used tools in research and industry to evaluate and validate networks, to study novel methods without the need for real deployments when resources are limited. However, existing network simulators do not support aging/degradation and reliability modeling or analysis. Popular network simulators, e.g., ns-3 [6], OMNeT++ [5], and OPNET [9] are equipped with a rich collection of communication models, allowing the assessment of network performance (throughput, delay, utilization, etc.) under different protocols. Recent research also integrated energy models and an energy-harvesting framework to the platform [25,27]. Yet it is not possible to analyze the reliability of IoT networks with the existing tools because they lack built-in reliability models. It should be noted that we refer to the aging and reliability degradation of the hardware of an IoT device, and not to the communication reliability.

To address this gap in reliability analysis, we propose a simulation framework named *RelIoT*, which allows practical and large-scale reliability evaluation of IoT networks. The framework is implemented in ns-3 [6], a discrete-event network simulator with low computational overhead and low memory demands. Up to one billion nodes can be simulated with ns-3 [19]. In recent years, with the addition of models for various network settings and protocols through open-source contributions, ns-3 has established itself as a de facto standard network simulation tool. To allow reliability simulation in ns-3, *RelIoT* integrates the following modules:

- *Power Module:* Supports power consumption simulation for various workloads with configurable power models.
- *Performance Module:* Works in cooperation with the power module to provide performance predictions for a given workload.

- *Temperature Module:* Estimates the internal temperature of a device based on its power consumption.
- *Reliability Module:* Evaluates the device reliability using the existing thermal-based degradation models [16, 23, 32].

To the best of our knowledge, *RelIoT*[1] is the first reliability analysis framework for heterogeneous IoT networks, taking thermal characteristics as well as power and performance into account. *RelIoT* enables researchers to explore trade-offs between power, performance, and reliability of network devices. Moreover, *RelIoT* incurs only a marginal performance overhead on the default ns-3, making it scalable for simulating large networks. For the scalability analysis of the default ns-3, the reader is referred to previous works [19, 20]. We validate our framework with two real-world experiments, showing *RelIoT* estimates power, performance, and temperature with errors of less than 3.8%, 4.5%, and $\pm 1.5\,^\circ$C respectively. We validate reliability models against the results from existing literature. Finally, we built a mesh network testbed to illustrate that *RelIoT* can effectively capture the average long-term power and thermal behavior of devices in a dynamic network. Finally, we provide example simulation results from *RelIoT* to motivate the need for reliability-aware management and to show the differences between energy-driven and reliability-driven management strategies.

The rest of the paper is organized as follows: Sect. 2 reviews power, performance, and reliability simulation techniques introduced by previous works. The overall structure of our proposed framework and details of models is elaborated in Sect. 3 and Sect. 4. Section 5 describes the evaluation setup and further discusses the results. The paper concludes in Sect. 6.

2 Related Work

Network Simulators for Power and Performance. Network simulators are used to study the behavior of computer networks and evaluate communication protocols prior to deployment. Popular examples are: ns-3 [6], OMNeT++ [5], and OPNET [9], all of which are discrete event-based and open-source. The standard versions of these network simulators are designed only for analyzing communication performance, lacking consideration for computation power, performance, and reliability of network devices.

Motivated by energy constraints in battery-powered sensor networks, several works have integrated power modeling and analysis with different granularity. Wu *et al.* [27] first introduced energy source models and device energy models to ns-3. They used existing analytical battery models and relied on hardware datasheets to build WiFi radio energy models. In another work named PASES [18], the authors construct accurate power consumption models for both processor and radio components of network devices by hardware design space

---

[1] *RelIoT* is available at: https://github.com/UCSD-SEELab/RelIoT.

exploration. In network simulators, usually, the accuracy of power models is compromised for attaining low computational costs. To provide flexible options for heterogeneous devices in an IoT network, *RelIoT* offers two configurable power models with different granularity, while allowing extensions for user-specified models.

For IoT performance, iFogSim [13] and EdgeCloudSim [22] incorporated computation performance (e.g. processing delay) to simulate end-to-end latency of a multi-level IoT structure including cloud servers, gateways, and sensors. Both toolkits employ low accuracy estimations such as look-up tables for latency and power analysis. *RelIoT* does a finer-grained estimation for different types of IoT devices running various applications.

Reliability Modeling and Management. Prior work has studied reliability degradation phenomena on processor-based systems. The considered failure mechanisms include Time-Dependent Dielectric Breakdown (TDDB), Negative Bias Temperature Instability (NBTI) and Electromigration (EM), which all limit device lifetime [16,23,32]. In these works, the reliability degradation problem is approached in two steps: (i) Physical-level models are built to quantify the reliability degradation due to voltage and temperature stress, which are influenced by the environmental conditions and workload variations. (ii) Based on the reliability degradation models, a management algorithm is designed to optimize performance while satisfying reliability constraints. The trade-off between performance and reliability could be adjusted during runtime by voltage scaling [16,23,32], task scheduling [10], or both [17]. The recent work by Mercati *et al.* [17] implements the above-mentioned models on a mobile phone, showing as much as a one-year improvement on lifetime with dynamic reliability management. Despite the impressive results on individual devices, reliability management for IoT networks is yet to be investigated. Recently, a dynamic optimization approach was proposed to manage battery reliability degradation in IoT networks, but their work does not consider the reliability of other device components (e.g., processor) [12].

In this work, we propose *RelIoT*, a framework for end-to-end reliability simulation in IoT networks to enable investigation of reliability trade-offs and prototyping of reliability management strategies. We develop and integrate power, performance, temperature, and reliability modules into ns-3. In contrast to prior work on network simulators, *RelIoT* offers temperature and reliability estimation for the networked devices.

3   Reliability Framework for RelIoT

In this section, we give a background on ns-3 and its features, then describe the overall structure of *RelIoT* and its integration with ns-3.

3.1   ns-3 Preliminary

ns-3 is built as a system of libraries that work together to simulate a computer network. To do simulation using ns-3, the user writes a *C++* program that links the various elements from the library needed to describe the communication network being simulated. ns-3 has a library of *objects* for all of the various elements that comprise a network (*objects* are highlighted in italics). *Nodes* are a representation of computing devices that connect to a network. Sensors, routers, hubs, gateways, and servers in the IoT architecture can be all considered *Node* objects. Figure 1a shows the structure of a typical ns-3 *Node*. *Net Devices* represent the physical device that connects a *Node* to a communications *Channel*. For example, the *Net Device* can be a simple Ethernet network interface card or a wireless IEEE 802.11 device, and the *Channel* could be a fiber–optic link or the wireless spectrum. *Packets* are the fundamental unit of information exchange in a network. A *Packet* contains headers describing the information needed by the protocol implementation and a payload which represents the actual data being communicated between network devices. Each protocol in the *Protocol Stack* performs some operation on network packets and then passes them to another layer in the stack for additional processing. The *Net Applications* are simple networking applications that specify the attributes of communication policies between devices. All of these individual components are aggregated on the *Node* objects

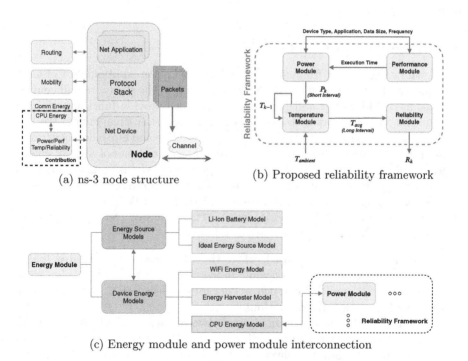

(a) ns-3 node structure

(b) Proposed reliability framework

(c) Energy module and power module interconnection

Fig. 1. ns-3 integration of the reliability framework

to give them communication ability and set up networking activity. Other modules, such as *Routing, Mobility,* and *Energy,* can be installed to provide additional functionality to *Nodes.*

3.2   Overview of the Proposed Framework

Our proposed framework consists of separate modules for power, performance, temperature, and reliability, as shown in Fig. 1b. IoT devices can run some applications to process the sensed or collected data before sending it to a central entity. In this case, as soon as an application starts, the performance module first calculates its execution time. Then, the power module gives an estimate of power consumption within the execution interval of the application. If the IoT device is not running any applications, then idle power consumption is estimated. Given the power estimation, ambient temperature, the temperature module outputs an estimated temperature, which is fed to the reliability module. Finally, reliability is calculated based on temperature. The modules operate on two different time scales: *Long Intervals,* on the order of days that it takes for reliability to change, and *Short Intervals,* on the order of milliseconds. Both performance and power values are updated every *Short Interval.* Reliability estimation is computationally expensive, so it is only done once every *Long Interval* using the average temperature over each interval. The underlying mechanisms of each module are be explained in Sect. 4.

3.3   Integration with ns-3

As shown in Fig. 1a, our framework is implemented as an additional set of modules that can be aggregated on the *Nodes,* adhering to the structure and conventions of ns-3. The power and performance modules provide functions to other modules for querying power consumption and execution time values. We also provide an interface connecting the ns-3 energy module and our reliability framework (Fig. 1c). The energy module (proposed in [27]) consists of a set of energy sources and device energy models. An *EnergySourceModel* is an abstraction for the power supply (e.g. battery) of a *Node.* The *DeviceEnergyModels* represent energy consuming components of a *Node,* for example, a WiFi radio. We implement a model called *CPUEnergyModel* as a child class of *DeviceEnergyModel.* The features of *CPUEnergyModel* are as follows:

- It is designed to be state-based, where the CPU can take the states *Idle* or *Busy.* The CPU will be *Busy* while processing packets received, e.g., while executing some applications such as encryption, decryption, compression, or Machine Learning (ML) algorithms.
- To determine when a transition occurs between states, a *PhyListener* is used. In ns-3, *PhyListener* is an object that monitors the network packet transmissions and receptions at the physical (PHY) layer. After an *Idle* node completes receiving data of specified size, the *PhyListener* notifies *CPUEnergyModel.* Then, the specified application is executed and state is set to *Busy.*

– It calculates the total energy consumed according to the power consumption value acquired from the power module and the execution time value acquired from the performance module. Then, it updates the remaining energy of the energy source described by *EnergySourceModel*.

4 Modules and Device Behavior Modeling

In this section, we describe the functionality of the proposed modules and interfaces and present underlying models in detail.

4.1 Power Module

The power module supports functions for running and terminating an application and switching between CPU states *Idle* and *Busy*. The value of power consumption is updated at a predefined period *Short Interval*, according to the selected power model. The power and temperature modules are interconnected; power consumption updates subsequently lead to temperature updates.

From cycle-accurate, instruction-level analysis to functional-level analysis, there are numerous power modeling techniques at different levels of abstraction. Low-level models use a fine-grain representation of the CPU, which usually implies that the time required for power estimation is large due to high computational complexity. This is undesirable for network simulations because it becomes very time consuming to simulate networks with a great number of nodes. In our framework, we offer two CPU power models having low model complexity while still providing good estimation accuracy. To improve the extensibility of the simulator for custom applications, we have included functionality for users to add new models to the power module. Parameters of the power models are configurable through external interfaces.

Frequency & Utilization-Based Power Model. The idea of estimating CPU power consumption on embedded devices based on CPU frequency and utilization is well studied. In a previous work [31], the authors use a linear combination of frequency and utilization to characterize the CPU power of a smartphone, achieving less than 2.5% average error. Similarly, we use linear models in our simulator to predict CPU power consumption P_{CPU}. The equation is given as:

$$P_{CPU}(t) = a \cdot f(t) + b \cdot u(t) + c \tag{1}$$

where $f(t)$ and $u(t)$ are CPU frequency and utilization at time t respectively. The coefficients a, b, c are learned through linear regression based on datasets collected on real devices. The frequency & utilization-based power model provides a good estimation accuracy for CPU power estimation on embedded devices. However, it requires frequency and utilization traces as inputs to the simulator which might not be available in practice.

Application-Based Power Model. The power consumption of embedded devices varies depending on the running application. An application-based model is convenient when there are only high-level functional properties are available, e.g.

input data size. Different applications have different power trends (i.e. linear, exponential, etc.) as the size of input data increases. Furthermore, the power consumed by running the same applications varies for different devices. In our framework, we adopt the modeling methodology proposed in [11], where the authors characterize and verify power models of running ML algorithms on edge devices (i.e. Raspberry Pi) and servers. They train, test, and cross-validate four regression models (linear, polynomial, log, and exponential regression), and select the best one. The input is the size of processed data by the application and the output is power consumption for these models. We leverage this methodology to deploy models for Raspberry Pi's and servers, but also apply the same methodology to build our own models for microcontrollers such as Arduinos. In addition to the 22 ML algorithms modeled in [11], our framework delivers a CPU power model for Multilayer Perceptron (MLP) based on the number of MAC (multiply-accumulate) operations. The same modeling approach can be applied to other neural network architectures such as Convolutional Neural Networks (CNNs).

4.2   Performance Module

IoT systems usually need to satisfy some performance requirements to provide adequate Quality of Service (QoS). To evaluate and monitor the performance of deployed applications and hence the overall network, we implement a performance module. Various metrics can be used to quantify performance, e.g., throughput, response time, etc. The performance metric is application-specific. For example, delay and throughput are critical in multimedia streaming applications whereas information accuracy is the main criterion for performance in some ML applications.

In our current release, we provide an *Execution Time Model*. We use the input data size of the application or number of MAC operations it needs to perform to estimate the application execution time. To build the model, we measure the execution times of various applications on a target device, then fit regression models to the collected data. Certain performance metrics can be calculated using the execution time value. For example, let t_{exec} be the execution time of an application, then its throughput can be obtained as D/t_{exec} where D is the input data size. In addition, end-to-end delay of a network path can be computed as the sum of communication and computation delays among the path (communication delay can be obtained using default ns-3 modules).

For both the power and performance modules, users are able to configure coefficients of the existing model or add new models with provided APIs.

4.3   Temperature Module

The goal of the temperature module is to estimate CPU temperature (based on CPU power consumption and ambient temperature) and to calculate the average temperature over a *Long Interval*. We adopt a thermal modeling strategy that can be used for any IoT device.

We assume that we do not have knowledge about the information describing topological and physical parameters of the device (e.g., we do not know material characteristics and layers of the devices' PCB board) so we cannot do a physical simulation of the process. To have an acceptable level of complexity in our simulator, we work on high-level information gathered from the coarse-grained thermal sensors of the device's key heat sources. Such information is available in most of the devices today like smartphones and single-board computers (e.g., Raspberry Pi).

Let the number of the heat sources be n and let $T_k \in \mathbb{R}^n$ represent the vector of temperatures observed by thermal sensors and $P_k \in \mathbb{R}^n$ be the power consumed by the heat sources at time instant k. Each heat source is assumed to have one thermal sensor measuring its temperature. Then, temperature T_{k+1} at time instant $k + 1$ can be predicted given the current temperature T_k and power P_k at time k. The discrete-time state-space model of the device's thermal behavior is expressed in Eq. (2) [8].

$$T_{k+1} = A \cdot T_k + B \cdot P_k + C \cdot T_k^{env} \tag{2}$$

where $A, B \in \mathbb{R}^{n \times n}$ are defined as the state and the input matrices. T_k^{env} is the ambient temperature and C is a vector of coefficients which weighs the impact of ambient temperature on each heat source's internal temperature. We use system identification methods to derive the model from measured power and temperature traces. A, B and C parameters are different for each class of devices, so we offer multiple device thermal models and made the parameters configurable through the temperature module API. The order of the model is equal to the number of the heat sources n. In our initial work, we use $n = 1$, where the only source is CPU. However, the extension to multiple sources is straightforward in our framework. For example, if a power model for GPU is provided, then power consumption values from both CPU and GPU can be used to predict temperature.

The temperature module updates the states in Eq. (2) at a time resolution of *Short Interval*, the same time granularity as power estimation updates. On the other hand, average temperature $\bar{T}$ is calculated for every *Long Interval* denoted *LI*. $\bar{T}$ is the exponential moving average of past temperature values in the interval k to $k + LI$.

$$\bar{T}_{k+1} = \alpha \cdot T_k - (1 - \alpha) \cdot \bar{T}_{LI} \tag{3}$$

where α is a weighing coefficient that is configured depending on the length of interval *LI*.

4.4 Reliability Module

The reliability module is the last component in the power, temperature, reliability module hierarchy. Temperature is estimated using power, while reliability is estimated using temperature. Unlike power and temperature, reliability is a slowly changing variable. Therefore, reliability can be estimated on a longer time scale, on the order of hours or days. Reliability degradation is affected more by average stress over a long time interval rather than instantaneous stress. We leverage these properties to calculate reliability sparsely because reliability models are highly compute-intensive. The reliability module does estimation every *Long Interval*, using temperatures averaged over the interval. It polls the temperature module to fetch the average temperature $\bar{T}_{LI}$ every *LI*, then $\bar{T}_{LI}$ is reset to start a new averaging operation.

Reliability is defined as the probability of not having failures before a given time t. To obtain the overall reliability of a processor, the effects of different failure mechanisms should be combined. We use the sum-of-failure-rates model as in RAMP [23], which states that the processor is a series failure system; the first instance of a failure due to any mechanism causes the entire processor to fail. In our reliability model, the single device reliability is a product of the reliabilities due to different failure mechanisms such as Time Dependent Dielectric Breakdown (TDDB), Negative Bias Temperature Instability (NBTI), Hot Carrier Injection (HCI), Electromigration (EM) and Thermal Cycling (TC). These mechanisms all depend on thermals.

Time Dependent Dielectric Breakdown (TDDB) Reliability Model. The thin gate oxide layer in transistors introduces a risk of breakdown and shortening devices lifetime. Due to gate oxide degradation, which is a non-reversible mechanism with a cumulatively increasing impact, a breakdown occurs. The reliability of a single transistor i with oxide thickness x_i subject to oxide degradation can be expressed as [24]:

$$R_i(t) = e^{-a(\frac{t}{\gamma})^{\beta x_i}} \tag{4}$$

where t is the time-to-breakdown, a is the device area normalized with respect to the minimum area, and γ and β are respectively the scale parameter and shape parameter. The scale parameter γ represents the characteristic life, which is the time where 63.2% of devices fail, and it depends on voltage and temperature. The shape parameter β, instead, is a function of the critical defect density, which in turn depends on oxide thickness, temperature and applied voltage. $R(t)$ is a monotonically decreasing function with values in the range of $[0, 1]$ indicating the probability that the system will not fail.

The reliability of the entire chip R_C can be expressed as the product of single transistor reliabilities:

$$R_C(t) = \prod_{i=1}^{m} R_i(t) = e^{\sum_{i=1}^{m} -a_i(\frac{t}{\gamma_i})^{\beta_i x_i}} \tag{5}$$

m is the number of transistors on the chip which can be on the order of millions. Since different regions of the chip have similar temperatures, the complexity possessed by large m on the computation of Eq. (5) can be reduced by assuming the same scale and shape parameters over the chip [32].

The R_C expression in Eq. (5) assumes a constant temperature applied from time $t = 0$, thus it is only representative of static systems. To capture the dynamics of reliability under varying temperature, we discretize the time and calculate reliability at each time step as shown in Eq. (6). The temperature is assumed to be constant between discrete time steps.

$$R_k = R_{k-1} - \Big(R_C(t_{k-1}, T_{k-1,k}) - R_C(t_k, T_{k-1,k})\Big) \qquad (6)$$

In Eq. (6), k indicates the kth time instant and $T_{k-1,k}$ is the temperature experienced by the chip between the time instants $k-1$ and k. We set this interval between adjacent time steps as the *Long Interval* and let $T_{k-1,k}$ be equal to the average temperature $\bar{T}_{LI}$ of the corresponding *LI*.

The reliability module can work with any failure mechanism or combination of multiple mechanisms as long as the mechanism can be described by a function $R_C(t)$, as in Eq. (5). For example, the module can be extended to include NBTI and HCI if we describe the reliability functions associated with these mechanisms, respectively R_{NBTI} and R_{HCI}. Then, by the sum-of-failure-rates approach, the reliability module calculates the total system reliability as the product of the functions associated with the single mechanisms as $R_C(t) = R_{TDDB}(t) \cdot R_{NBTI}(t) \cdot R_{HCI}(t)$. Equation (6) would not need any modifications since it is general and does not depend on a specific $R_C(t)$.

5 Experiments and Results

In this section, we first present validation results on a three-node network topology, comparing power, performance, and temperature measurements from experiments with the simulated traces. We then use a testbed with a mesh network of 10 heterogeneous nodes to evaluate the accuracy of the simulator under different networking conditions and temperatures. We cannot explicitly validate reliability because it requires long term experiments and specialized degradation sensors. Finally, we illustrate how the proposed simulator is useful in exploring energy, performance, reliability trade-offs in a network and show that it can be used to implement reliability-aware strategies. We analyze examples of energy-optimized and reliability-optimized network configurations to motivate reliability-aware network design and management.

5.1 Validation and Evaluation

Three-Node Network Topology. To validate the device models and to verify the functionality of the simulator modules, we use a simple three-node network. The setup consists of an ESP8266 WiFi microchip with microcontroller, a Raspberry Pi 3 (RPi3), and a PC. The devices communicate over WiFi (IEEE 802.11b) and transmit/receive TCP/IP packets using MQTT protocol [4]. The ESP8266 samples random data as a sensor node, runs median filtering to preprocess the data, and sends filtered data to RPi3. The data is further processed by an application on the RPi3, or the computation can be offloaded to the PC. This type of computation offloading is common in IoT edge devices and is representative of their usual operation [21]. If the application is chosen to be offloaded, then the RPi3 is only responsible of relaying incoming data to the PC.

In our three-node experiments, we collected 5 different measurements synchronously:

(i) RPi3 power consumption (via HIOKI 3334 power meter [2]),
(ii) RPi3 CPU temperature (via built-in temperature sensor),
(iii) ESP8266 power consumption (via INA219 power monitor [3]),
(iv) ESP8266 CPU temperature (via built-in temperature sensor),
(v) Ambient temperature (via DHT22 temperature sensor [1]).

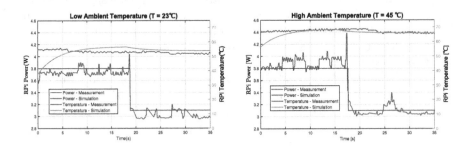

Fig. 2. RPi3 power and temperature traces

Measurements and simulation results are presented in Fig. 2 for an example test case under two different ambient temperatures. The goal here is to show a temporal view of the simulator output, particularly in a dynamic case where the simulated device has a varying workload. In this experiment, the RPi3 runs a data processing application with incoming data input from ESP8266 for the first 15–20 s. After that, the application is offloaded to the PC and the RPi3 only relays data while its CPU is *idle*. As shown in Fig. 2, the simulator output follows the real power and temperature traces with a mean error of 3.42% and 6.19% in low ambient temperature, and with a mean error of 2.69% and 3.97% in high ambient temperature. The discrepancy between real and simulated temperatures at the beginning of each plot is because of the initial condition set for the temperature in the simulator. The temperature starts from a lower initial condition and reaches a steady-state value.

Overall, applying the same modeling methodology of reference [11], we estimate the execution time and energy consumption of the RPi3 for 23 different ML applications with average errors of 3.8% and 4.5%, respectively. For the CPU temperature, the state-space model predictions stays within ±1.5 °C of measurements at steady-state, for all applications.

Mesh Network Topology. To show that our simulator can correctly capture devices' behavior in a more complicated scenario, we simulate a larger network under different configurations and operating conditions, then validate it using our testbed. As shown in Fig. 3, the testbed spans a whole floor in UCSD CSE

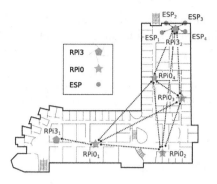

Fig. 3. Mesh network topology

department building, including two Raspberry Pi 3 (RPi3), four Raspberry Pi 0 (RPi0) and four ESP8266. Data is generated from each node and communicated to the sink node $RPi3_1$ in multiple hops via MQTT. The network of all RPis works in an ad-hoc manner, while all ESP8266s forward their data to $RPi3_2$ that is a gateway for that local area. Not all devices can communicate with each other because some pairs are out of communication range. The connections are depicted in Fig. 3. Using this setup, we both implement and simulate following scenarios:

Scenario 1. $RPi3_1$ and $RPi0_1$ process the data, while the other devices only communicate. The ambient temperature for each network device is approximately 25 °C.

Scenario 2. The same devices process data as Scenario 1. We use a heater that raises the ambient temperature around $RPi0_1$ to 37 °C, while the rest of the devices are in the normal ambient condition of 25 °C.

Scenario 3. The data processing duties of $RPi3_1$ and $RPi0_1$ are distributed between $RPi0_2$, $RPi0_3$, and $RPi0_4$. Therefore, each of these three devices only transmits the outputs of data processing tasks to $RPi0_1$, which directly forwards them to $RPi3_1$. $RPi0_1$ is still in a heated environment of 37 °C.

An ML application can be split and distributed to edge devices, which allows us to realize the different configurations in these scenarios for allocating data processing without changing the overall application behavior [26,28]. Figure 4 illustrate the power and temperature distribution of the devices, while Fig. 5 shows the simulated reliability traces in a year. We only depict the measurement and simulation statistics on $RPi0s$ in each scenario, but the rest show similar trends. Comparing collected traces to simulation logs, our result shows that *RelIoT* is able to estimate average power within ± 0.11 W ($\sim$11%), and average temperature within ± 4 °C ($\sim$9%). It can be seen from Fig. 4 that, although extremities in both power and temperature are difficult to track, *RelIoT* is able to precisely capture the averages in different configurations. Scenario 3 distributes the workload to other $RPi0s$, thus significantly reduces the network traffic. Consequently, power and temperature of both $RPi0_1$ and the rest $RPi0s$ drop, which is also reflected in the simulation *RelIoT* starts simulation from a device temperature of 35 °C, which explains why the minimum temperature of *RelIoT* consistently locates at 35 °C.

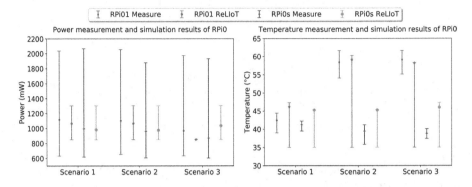

Fig. 4. Collected and simulated statistics of RPi0s (average, max, and min values).

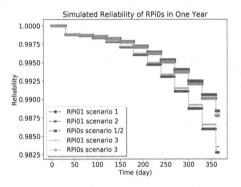

Fig. 5. Reliability degradation of RPi0s in one year.

The power and temperature validation experiments lasts 300 s, but we simulate the network for a time-span of one year to observe the long-term reliability changes. The stair pattern in Fig. 5 is a result of *RelIoT* updating the reliability by the end of each *Long Interval*. In all scenarios, reliability remains fairly high if the device is in normal ambient temperature. In Scenario 1, $RPi0_1$ degrades slightly faster than the rest of the $RPi0_s$ due to its data processing workload. However, in Scenario 2, the raised ambient temperature together with its workload lead to a drastic degradation in reliability. In such case, workload reallocation as in Scenario 3 can alleviate degradation. The network devices in environments with low temperatures can take on a higher workload to mitigate reliability problems of the quickly degrading devices. The result in Fig. 5 implies the necessity for reliability-aware management in IoT networks.

5.2   Reliability-Aware Management

Most of the IoT devices are battery-powered and/or rely on energy harvesters with limited energy sources. Therefore, traditionally, many network management solutions aim at optimizing the energy consumption while satisfying some Quality-of-Service (QoS) constraints (throughput, delay, jitter, network coverage, etc.). In this context, reliability is also a design parameter that can be optimized or a certain overall reliability constraint can be subjected to the network. Although correlated, the optimal energy efficiency and reliability usually are not ensured by the same management strategy. The designers need to find good trade-offs between energy savings and reliability. In this section, we show how our simulator addresses this issue by making reliability-aware management and design possible. To emphasize the differences between two approaches and to motivate reliability-aware management, we provide simulation results for different scenarios of energy-optimized and reliability-optimized network management strategies using the topology in Fig. 3.

Energy-Optimized. Our interest here is to partition an application into smaller tasks and find the task allocation that maximizes the lifetime of a network. Many ML applications can be partitioned while preserving functionality [26,28]. For each device in the network, the energy consumed for computing and communicating data of size s is given as:

$$\text{Computation:} \qquad P_{device}^{\xi}(s) \times t_{exec}(s) \qquad (7)$$

$$\text{Communication:} \qquad P_{wifi}(d, BW) \times \frac{s}{BW} \qquad (8)$$

where ξ denotes the application, t_{exec} is the application execution time, d denotes the communication distance, and BW is the communication bandwidth. $P_{wifi}(d, BW)$ is the power consumption of WiFi which can be parameterized by distance and bandwidth allocation BW [11]. In an energy-optimized application partition, the mapping of tasks to the devices depends on:

(i) Power characteristics of the application,
(ii) Execution time,
(iii) Allocated bandwidth,
(iv) Distance between the neighbouring devices.

We adopt the convex optimization formulation from [15] and apply it to our problem, with a slight modification by adding the computation energy term in Eq. (7). We find the optimal partitioning of the application such that the *maximum* energy consumption among network devices due to computation and communication of the data is minimized.

Reliability-Optimized. Similar to the previous case, we map the tasks of an application to the network devices. We use the same solution approach, but this time, the objective is to maximize the *minimum* reliability among network devices. Reliability of each device is $R_{C,device}(t, T)$, which is dependent on time and temperature. We simulate a time horizon t_{sim}, so we want to optimize for $R_{C,device}(t_{sim}, T)$. This is under the assumption of environment temperature T_{amb} being constant for the entire horizon. In the following experiments, a static solution (constant for the whole time horizon) is simulated for both energy-optimized and reliability-optimized cases, but it can be made dynamic by solving for the current energy and reliability estimates at each time instant, as in (6). In this way, the solution can adapt to changing network configurations (bandwidth, applications) and operating conditions (environment temperature).

Figure 6 presents the comparison of energy-optimized and reliability-optimized solutions for different bandwidth and environment temperature configurations. The network devices run a part of a data processing application where the optimal partitions are determined by the two approaches. The energy-optimized partition brings 1.0%, 9.1%, and 10.9% better energy efficiency

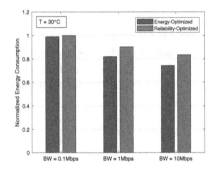 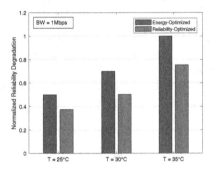

Fig. 6. Energy consumption and reliability degradation results for two approaches

compared to the reliability-optimized partition for 0.1 Mbps, 1 Mbps, and 10 Mbps bandwidth configurations respectively. Referring to Eq. (8), it can be seen that as bandwidth increases, the time it takes to communicate data of size s decreases, hence, decreasing the communication energy. The difference of energy efficiencies between two solutions are increasing with bandwidth because the energy-optimized solution leverages the decrease in communication energy and allocates more communication instead of computation to the higher energy consuming network devices. On the other hand, the reliability-optimized partition results in 25.0%, 28.2%, and 24.5% less reliability degradation compared to the energy-optimized partition for 25 °C, 30 °C, and 35 °C environment temperatures respectively. The reliability-optimized solution allocates less computation on the most degrading network devices, conserving reliability. These results show that, although being correlated, the optimal energy efficiency and reliability do not yield from the same management strategy. Therefore, if the concern is particularly the reliability, a reliability-aware management strategy should be adopted.

6   Conclusion

We presented a novel framework for the reliability analysis of IoT networks using the ns-3 network simulator. The proposed framework can be used to explore trade-offs between power, performance, and reliability of devices in a network. We validated our reliability framework in two experimental setups: a three-node network and a ten-node mesh network. Additionally, we motivated the need for reliability-aware management through example simulation results of energy-optimized and reliability-optimized management strategies. As future work, we plan to leverage our framework for design space exploration (DSE) of IoT networks. We can simulate, explore, and check the feasibility of different network configurations in terms of different objectives such as energy efficiency, reliability, and performance. We believe that our contribution will help researchers to study the reliability degradation problem in large-scale networks.

Acknowledgements. This work was supported in part by SRC task #2805.001, NSF grants #1911095, #1826967, #1730158 and #1527034, and by KACST.

References

1. DHT22 Datasheet. https://www.sparkfun.com/datasheets/Sensors/Temperature/DHT22.pdf/
2. Hioki3334 Powermeter. https://www.hioki.com/en/products/detail/?product_key=5812
3. INA219 Datasheet. http://www.ti.com/lit/ds/symlink/ina219.pdf/
4. MQTT MQ Telemetry Transport. https://mqtt.org/
5. OMNeT++ Discrete Event Simulator. https://omnetpp.org/
6. The ns-3 Network Simulator. https://www.nsnam.org/
7. The Hidden Costs of Delivering IIoT (2016). https://www.cisco.com/c/dam/m/en_ca/never-better/manufacture/pdfs/hidden-costs-of-delivering-iiot-services-white-paper.pdf
8. Beneventi, F., Bartolini, A., Tilli, A., Benini, L.: An effective gray-box identification procedure for multicore thermal modeling. IEEE Trans. Comput. **63**(5), 1097–1110 (2012)
9. Chang, X.: Network simulations with OPNET. In: WSC 1999. 1999 Winter Simulation Conference Proceedings. Simulation-A Bridge to the Future (Cat. No. 99CH37038), vol. 1, pp. 307–314. IEEE (1999)
10. Coskun, A.K., Rosing, T.S., Whisnant, K.: Temperature aware task scheduling in MPSoCs. In: 2007 Design, Automation & Test in Europe Conference & Exhibition. pp. 1–6. IEEE (2007)
11. Cui, W., Kim, Y., Rosing, T.S.: Cross-platform machine learning characterization for task allocation in IoT ecosystems. In: 2017 IEEE 7th Annual Computing and Communication Workshop and Conference (CCWC), pp. 1–7. IEEE (2017)
12. Ergun, K., Ayoub, R., Mercati, P., Rosing, T.: Dynamic optimization of battery health in IoT networks. In: 2019 IEEE 37th International Conference on Computer Design (ICCD), pp. 648–655, November 2019. https://doi.org/10.1109/ICCD46524.2019.00093

13. Gupta, H., Vahid Dastjerdi, A., Ghosh, S.K., Buyya, R.: iFogSim: a toolkit for modeling and simulation of resource management techniques in the Internet of Things, edge and fog computing environments. Softw. Pract. Exp. **47**(9), 1275–1296 (2017)

14. International Data Corporation: The Growth in Connected IoT Devices (2019). https://www.idc.com/getdoc.jsp?containerId=prUS45213219

15. Chang, J.-H., Tassiulas, L.: Maximum lifetime routing in wireless sensor networks. IEEE/ACM Trans. Netw. **12**(4), 609–619 (2004). https://doi.org/10.1109/TNET. 2004.833122

16. Karl, E., Blaauw, D., Sylvester, D., Mudge, T.: Reliability modeling and management in dynamic microprocessor-based systems. In: Proceedings of the 43rd annual Design Automation Conference, pp. 1057–1060. ACM (2006)

17. Mercati, P., Paterna, F., Bartolini, A., Benini, L., Rosing, T.Š.: WARM: workload-aware reliability management in Linux/Android. IEEE Trans. Comput. Aided Des. Integr. Circuits Syst. **36**(9), 1557–1570 (2016)

18. Minakov, I., Passerone, R.: PASES: an energy-aware design space exploration framework for wireless sensor networks. J. Syst. Arch. **59**(8), 626–642 (2013)

19. Nikolaev, S., Banks, E., Barnes, P.D., Jefferson, D.R., Smith, S.: Pushing the envelope in distributed ns-3 simulations: one billion nodes. In: Proceedings of the 2015 Workshop on Ns-3, WNS3 2015, pp. 67–74. Association for Computing Machinery, New York (2015). https://doi.org/10.1145/2756509.2756525. https://doi.org/10.1145/2756509.2756525

20. Nikolaev, S., et al.: Performance of distributed ns-3 network simulator. In: Proceedings of the 6th International ICST Conference on Simulation Tools and Techniques, SimuTools 2013, pp. 17–23. ICST (Institute for Computer Sciences, Social-Informatics and Telecommunications Engineering), Brussels, BEL (2013)

21. Samie, F., Tsoutsouras, V., Masouros, D., Bauer, L., Soudris, D., Henkel, J.: Fast operation mode selection for highly efficient IoT edge devices. IEEE Trans. Comput. Aided Des. Integr. Circuits Syst., 1 (2019). https://doi.org/10.1109/TCAD. 2019.2897633

22. Sonmez, C., Ozgovde, A., Ersoy, C.: EdgeCloudSim: an environment for performance evaluation of edge computing systems. Trans. Emerg. Telecommun. Technol. **29**(11), e3493 (2018)

23. Srinivasan, J., Adve, S.V., Bose, P., Rivers, J.A.: The case for lifetime reliability-aware microprocessors. In: ACM SIGARCH Computer Architecture News, vol. 32, p. 276. IEEE Computer Society (2004)

24. Stathis, J.H.: Physical and predictive models of ultrathin oxide reliability in CMOS devices and circuits. IEEE Trans. Device Mater. Reliab. **1**(1), 43–59 (2001)

25. Tapparello, C., Ayatollahi, H., Heinzelman, W.: Energy harvesting framework for network simulator 3 (ns-3). In: Proceedings of the 2nd International Workshop on Energy Neutral Sensing Systems, pp. 37–42. ACM (2014)

26. Thomas, A., Guo, Y., Kim, Y., Aksanli, B., Kumar, A., Rosing, T.S.: Hierarchical and distributed machine learning inference beyond the edge. In: 2019 IEEE 16th International Conference on Networking, Sensing and Control (ICNSC), pp. 18–23, May 2019. https://doi.org/10.1109/ICNSC.2019.8743164

27. Wu, H., Nabar, S., Poovendran, R.: An energy framework for the network simulator 3 (ns-3). In: Proceedings of the 4th International ICST Conference on Simulation Tools and Techniques, pp. 222–230. ICST (Institute for Computer Sciences, Social-Informatics and Telecommunications) (2011)

28. Yao, S., Zhao, Y., Zhang, A., Su, L., Abdelzaher, T.: DeepIoT: compressing deep neural network structures for sensing systems with a compressor-critic framework. In: Proceedings of the 15th ACM Conference on Embedded Network Sensor Systems, pp. 1–14 (2017)
29. Zhai, D., Zhang, R., Cai, L., Li, B., Jiang, Y.: Energy-efficient user scheduling and power allocation for NOMA-based wireless networks with massive IoT devices. IEEE Internet Things J. **5**(3), 1857–1868 (2018)
30. Zhang, H., Xiao, Y., Bu, S., Niyato, D., Yu, F.R., Han, Z.: Computing resource allocation in three-tier IoT fog networks: a joint optimization approach combining Stackelberg game and matching. IEEE Internet Things J. **4**(5), 1204–1215 (2017)
31. Zhang, L., et al.: Accurate online power estimation and automatic battery behavior based power model generation for smartphones. In: Proceedings of the Eighth IEEE/ACM/IFIP International Conference on Hardware/Software Codesign and System Synthesis, pp. 105–114. ACM (2010)
32. Zhuo, C., Sylvester, D., Blaauw, D.: Process variation and temperature-aware reliability management. In: Proceedings of the Conference on Design, Automation and Test in Europe, pp. 580–585. European Design and Automation Association (2010)

NACK-Based Reliable Multicast Communication for Internet of Things Firmware Update

Jiye Park[1(✉)], Dongha Lee[2], Markus Jung[3], and Erwin P. Rathgeb[1]

[1] University of Duisburg-Essen, Essen, Germany
{ji-ye.park,erwin.rathgeb}@uni-due.de
[2] Munich, Germany
mathicalee@gmail.com
[3] OSRAM GmbH, Munich, Germany
m.jung@osram.com

Abstract. The demand for efficient IoT firmware update protocols is increasing. Especially in scenarios with a large number of constrained devices, transferring a big amount of data, like a firmware image file, over a constraint network takes a long time to complete. During this time the functionality of the devices may be reduced. Therefore the firmware update is a critical use case for IoT. Multicast group communication can shorten the transmission time and use the network bandwidth efficiently. However, current IoT protocols using multicast cannot guarantee reliability which is most important for firmware file transmission. Furthermore existing solutions for reliable multicast cause a significant network overhead which can be prohibitive for a constrained IoT environment. To address these problems, we propose a reliable multicast solution employing a Negative ACK (NACK) mechanism that can be integrated with the Constraint Application Protocol (CoAP) widely used in IoT. Our protocol mitigates network congestion by reducing the number of packets that have to be sent while keeping the message size small making it a suitable solution even with a large number of devices. What is more, our proposal does not require an additional stack. In order to demonstrate the feasibility and effectiveness of our proposal, we carried out a real-world evaluation in a wireless mesh network testbed.

Keywords: Internet of Things · Firmware update · Negative ACK · Multicast communication · CoAP · Block-wise

1 Introduction

In the Internet of Things (IoT), most of the devices have constrained hardware resources (i.e. limited size of RAM, flash and slow microprocessor) and a constrained network such as IEEE 802.15.4 having 127 bytes MTU size. Those devices are meant to do simple communication such as sending a small amount of diagnostic data to the cloud and receiving a simple control message.

© Springer Nature Switzerland AG 2020
W. Song et al. (Eds.): ICIOT 2020, LNCS 12405, pp. 82–95, 2020.
https://doi.org/10.1007/978-3-030-59615-6_6

Nevertheless, firmware update for those devices is very important not only to update features and configurations but also to keep the devices up to date in a security aspect. Considering the devices are not replaced within a short time once they are deployed, firmware update is prolonging device life cycle and makes the IoT system secure.

The firmware update process is an exceptional case that the constrained IoT devices have to handle a large amount of data including a large number of packets being transmitted for a relative long duration in Low power and Lossy Network (LLN). When the large amount of data is transferred, the data is fragmented on the IP adaptation layer in order to fit in the IEEE 802.15.4 MTU [7]. This fragmentation causes significant overhead in the constrained network since one fragmented packet is lost, there is no way to figure out which part of data is lost on the application layer and thus all data has to be re-transmitted. To avoid the fragmentation, the block-wise option [2] is standardized as an extension of basic CoAP. Even with the block-wise option, transferring big firmware file to individual device takes very long especially when there are hundreds of devices to update.

To shorten the total transmission time and to use the network bandwidth efficiently, multicast can be used when big amount of data should be transferred to many devices in the same network. However, the current CoAP standard specifies not to use the confirmable message type for multicast messages due to the congestion control. Thus it cannot provide reliability which is very important for firmware file transmission.

In this paper, we present a Negative ACKnowledgement (NACK) approach which is mitigating network congestion. By utilizing CoAP option field, our proposal can be integrated with CoAP to realize reliable multicast for constrained IoT environments. We consider a realistic IoT scenario involving constrained devices deployed in a wireless mesh network. By setting up a testbed, we check the feasibility of firmware updates through multicast and benchmark it against a unicast approach. The results present a comparison between a reliable update mechanism using multicast and a unicast-based communication approach.

This paper is organized as follows. In Sect. 2, we discuss about existing standard protocols and related research papers. In Sect. 3, we propose Negative type (Nety) CoAP option. Section 4 analyses our proposal based on three different scenarios to support reliable multicast communication. In the following Sect. 5, we evaluate our proposal. Finally, Sect. 6 presents our conclusions.

2   Related Works

2.1   Related Standard Protocols

Related protocols with respect to the IoT firmware update use case are discussed below:

- **Constrained Application Protocol (CoAP)** is a RESTful application protocol running over UDP for constrained IoT devices [11]. To take benefit of using UDP and support reliability at the same time, CoAP defines two different message types: Confirmable (CON) and Non-confirmable (NON).

With CON type in the CoAP header, sender tries maximum three times data retransmission in case Acknowledgement (ACK) message has not arrived within a certain time. To avoid network congestion caused by the many ACKs from receivers, in a multicast communication scenario, only NON type messages shall be sent and reliability is not guaranteed [9].

– **Block-wise** was proposed as a CoAP option to segment large data to blocks having suitable size on the application layer to avoid IP layer fragmentation [2]. Depending on a CoAP message where payload is placed either request (PUT, POST) or response (GET), block option 1 or block option 2 is chosen. Block wise option has three fields: Number (NUM), More (M) and Size. The NUM field is used to show the sequence number of the current block and the M field indicates if there are more blocks that will be delivered. The Size field is showing the current block size. In a multicast firmware update scenario, firmware file blocks are placed as a payload in request messages with PUT or POST method. Therefore block wise option 1 should be used, but the specification does not define how to handle multicast packets with Block option 1 yet.

– **Multicast protocol for low power and lossy network (MPL)** provides efficient multicast routing mechanism in constrained network based on the trickle algorithm [5]. The protocol is using IPv6 header option field for the routing. Since the protocol is robust and scalable, it is used in the OpenThread[1] stack. However the protocol does not guarantee reliability.

– **Negative ACK Oriented Reliable Multicast (NORM)** is a well designed protocol for supporting reliable multicast communication by leveraging NACK and Forward Error Correction-based (FEC) repair [1]. Nonetheless, using this protocol in a constrained IoT environment is not the best idea considering that it is designed for non-constrained Internet environments. The protocol is placed in between application layer and transport layer. Thus, an additional layer in the IoT stack is required. Flashing an IoT firmware with large size of additional library is not possible for general memory constrained IoT devices. Besides, given that the CoAP header size is only 4 bytes, additional 32 bytes for the NORM header is a burden for the lossy network having the MTU limitation. Supposing the NORM protocol is used with CoAP, many header fields are even duplicated. Therefore the protocol is difficult to be applied as it is. To use NORM in IoT, protocol optimization and modification are necessary.

2.2 Firmware Update for IoT

The difficulties of IoT firmware updates are addressed in many papers [3,4,6,10]. Among them, [10] showed limitations of using IoT standard protocols to transmit large size of data and proposed an error tolerant network coding extension for CoAP to increase efficiency by reducing the loss-indicated problems. Especially the paper [6] points out that updating one device at a time is not feasible

---

[1] https://www.threadgroup.org/support#Whitepapers.

approach in the IEEE 802.15.4 network. In that point, the proposal separated steps for firmware file dissemination from a server to controllers. The proposal is leveraging the Deluge algorithm for the last step to transmit the file to the controllers.

[8] introduced CoNORM architecture providing reliable multicast communication. The proposal implemented NORM connector in Java based CoAP library, Californium (cf) to add the NORM layer and evaluated performance using a Raspberry Pi setup. The considered environment of this paper is dynamic and unpredictable MANET whereas we focus on constrained IoT environments. The paper did not consider the header fields that become redundant by having both CoAP and NORM protocols. Therefore, the overhead issue caused by the additional header is still remained to use the proposal in constrained network.

3   Negative Type Option for CoAP

In the previous section we have seen that the NORM protocol implements a NACK-scheme in order to provide reliability by introducing a new protocol layer (between the transport and application layer) causing additional overhead which is prohibitive for IoT. Our approach is to take the CoAP protocol and implement a NACK-type mechanism, not by introducing a new protocol layer but directly within the application layer. Our proposal, the Negative type (Nety) CoAP option, provides the means to enable reliable multicast in CoAP.

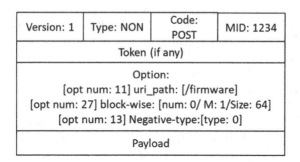

Fig. 1. CoAP message format with negative type option

The CoAP message is transmitted in the format as described in Fig. 1 from a sender (firmware file server) to receivers (IoT devices) having the same multicast ip address. The Message ID (MID) represents the message identifier, and it increases sequentially. To avoid IP layer fragmentation we use the block-wise option together with the Nety option. In order to include Nety in CoAP, we use the CoAP option number 13 not yet assigned by IANA. We also assume that the block size among the communication peers can be negotiated in advance.

Figure 2 shows the Nety option fields. The Nety option uses one bit of the type field to present two states: Negative CONfirmable (NCON) and NACK. NCON

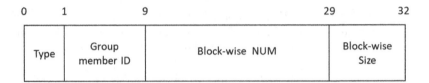

Fig. 2. Negative ACK type option value

has type value 0. When the type value is set to 0 in the option, the option works as a flag to let the receivers know that the sender requests a NACK message in case packet loss is detected. In case of the NCON type, remaining Nety option fields: group member ID, block-wise NUM and block-wise size are all set to 0. According to CoAP encoding rules, a zero byte integer is sent for each field that is set to 0 so that packet size can remain small. Even if the CoAP type header is set to NON, Nety type 0 allows the receiver to send NACK message to the sender when a packet loss happens.

In the Nety option, the type value 1 indicates the NACK message. The NACK message is transmitted from a receiver to a sender while the receiver is receiving blocks having the NCON type of Nety option and detects packet loss. In contrast to the NCON type where all the fields are set to 0, the NACK type has values of following fields.

To ask a sender for retransmission of the particular packet, the receiver adds 1) its group member ID to be distinguished from other receivers, 2) the block number of the packet required to be retransmitted and 3) the block size information. For the group member ID field, 8 bits are allocated so in total 256 devices can be covered in one multicast group. 20 bits of the block-wise num field can cover the case that a firmware file is a bit bigger than 1 MB, and it is transmitted with the smallest block size which is 16 bytes. The block size field has 3 bits as it is standardized in [2], and the information is used for the sender to find the correct block to re-transmit. Still, the NACK type has at maximum 4 more bytes overhead. Since a NACK does not carry any payload, it is not a burden to a constrained network. The message having NACK type option is transmitted as a new request with CON type in the CoAP header over unicast, not a response message.

4   Protocol Description and Packet Loss Scenarios

Unlike CON-type based communication where the sender is taking care for the data loss detection and retransmission, in the Nety-based communication, the sender does not retransmit a packet unless the receiver detects the packet loss and requests the data retransmission. Thus receivers are expected to do those tasks in our proposal. To allow this, the receivers need to know when the communication starts and ends. In this section we explain how the receivers detect the data loss when the block-wise and Nety option are used for a reliable multicast using three different scenarios. In all the scenarios, the block size is set to 64 byte to avoid

IP layer fragmentation. We assume that the first block always contains manifest of the firmware file and information about the total number of blocks that will be transferred.

4.1   Scenario A: The First Packet Is Lost

To avoid a circumstance that CoAP messages having the same MID from different senders are considered as a duplicated message, the MID is determined randomly when the CoAP client prepares the first message, and the MID is increased sequentially until the end of the communication. Therefore it is difficult that a receiver figures out if the packet is the first one or not because the receiver does not have the previous packet to compare the sequence of the MID. In this case the server must check the NUM field of block-wise option. If the NUM field is not set to 0, the server considers the first packet as lost and sends a NACK message to the client with the CON type header in a unicast message.

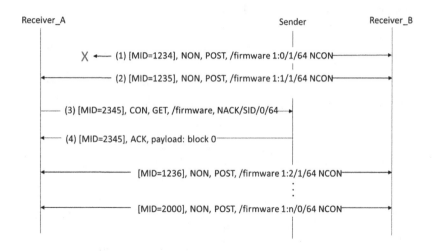

Fig. 3. Message flow when the first packet is lost

Figure 3 shows messages that are exchanged between sender and receivers when the first packet is lost. All the packets transmitted from the sender to receivers with block-wise option have the NON type in the CoAP header and the NCON type in the Nety option field. In this scenario, the first block (1) is lost, and the receiver A did not receive it. The receiver receives blocks only from the block (2). When the block (2) arrives at receiver A, it knows that there was no previous blocks transmitted before the block (2) by comparing the MID of current block from the same sender to previous received blocks. As a next step, the receiver checks if the block (2) is the first block or already the first block is lost. In this case the receiver checks block number not the MID. As the block number of the message (2) is 1 not 0, the receiver can determine that the first block having block number 0 is lost. Then the receiver sends a retransmission request message (3) including NACK type value, the receiver's group member id and the request block information in the Nety option. The message (3) is transmitted in unicast with CON type in CoAP header. The sender who receives the message (3) checks the option field to know which block is requested, and it sends piggyback response that the requested block is placed in the payload of the ACK response message (4).

4.2   Scenario B: A Middle Packet Is Lost

Once the first block is received well, detecting packet loss for the rest of the blocks is relatively easy. Figure 4 describes a scenario in which a packet loss happens in the middle of the transmission and the retransmitted packet is also lost. The first packet (1) has MID 1234 which is decided by a sender with NON type header. The packet includes block-wise option set to block 1 and block number 0, and Nety option. MID and block number are sequentially increased in the following packets (2), (3) and (4).

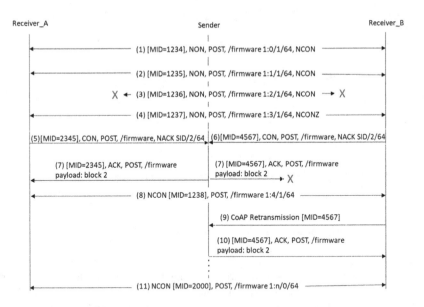

Fig. 4. Message flow when a middle packet is lost

On the receiver side MID and block number are checked every time a packet arrives. When the packet (4) arrives, the receiver notices that packet (3) was not received by comparing the MID in the packet (4) to the MID in the previous packet (2). Then the receiver sends the retransmission request message as described in previous scenario A. In the packet, the type of the Nety option value is set to 1 and block-wise num field has the block number of the missing packet (3). Since the NACK message is not a response but a request, and the receiver sends it as a CoAP client, the MID is randomly decided by the receiver. The sender receiving the unicast message sends the piggyback responses (5, 6) including the requested block payload. In case the retransmitted packet is lost again (8), the receiver B retransmits the message (7) again to the sender after the CoAP timer expires. The retransmission process is following the CoAP retransmission process since the message is sent with the CON type header in unicast.

4.3   Scenario C: The Last Packet Is Lost

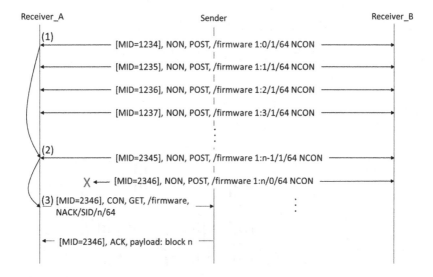

Fig. 5. Message flow when the last packet is lost

As the block-wise number is counted from zero, the total number of the blocks is equal to *block-wise number+1*. From the first message (1) in the Fig. 5, the receiver knows the total number of blocks. A timer is started when the latest packet received has the block number *total number of blocks-2*. If the last packet to receive, having the block number equal to the total *number of the blocks-1* and a zero value in the M field, is not transferred until the timer expires, the receiver sends a retransmission request message including the Nety option. The response to the NACK message is sent as a piggyback response as described above. The retransmitted block in the payload has zero value in the M field. When the receiver receives a block having the zero value in M field in block-wise option, it considers all data blocks are transmitted and communication is finished.

5 Evaluations

5.1 Calculation of Total Number of Packets

To compare the total number of packets that have to be sent to all the IoT devices to have the new firmware file, we calculated it in three different scenarios, (1) unicast with confirmable message (UwC) (2) multicast with confirmable message (MwC) (3) multicast with Nety option (MwN), considering different number of devices and different packet loss rate (PLR). When the number of original blocks is B, the number of devices is D and packet loss rate is PLR, the total number of packets P per scenario case is following the formula as below.

$$P = B * D + (1 - PLR) * B * D + 2 * B * PLR * D \tag{1}$$

$$P = B + (1 - PLR) * B * D + 2 * b * PLR * D \tag{2}$$

$$P = B + PLR * D + 2 * B * PLR * D \tag{3}$$

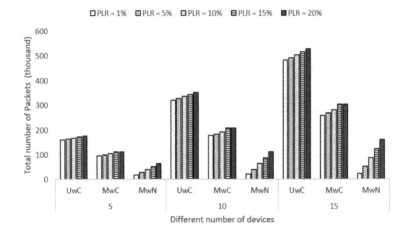

Fig. 6. Total number of packets comparison

Figure 6 shows the comparison of the total number of packets when a 100 KB firmware file is transmitted in 64 byte size of blocks where the number of deployed devices is 5, 10 and 15 with different PLR. In case of multicast with the Nety option, the number of additional packets for providing reliable communication is significantly lower compared to the unicast based approach. Our approach can transmit a large amount of data efficiently in any case of PLR with the lowest number of additional packet overhead compared to unicast with CON and multicast with CON scenarios. Especially when there are more devices deployed in a network with a low PLR, our solution can be more efficient.

5.2   Experimental Results

To evaluate our proposal in the real field, we set up a testbed in an office building. We used the nRF52840 dongle (PCA 10056) as an IoT end device (see Fig. 7) and a Raspberry Pi as a border router. Table 1 shows the specification of the IoT device we used. For the Network Co-Processor (NCP) allowing the border router to have an IEEE 802.15.4 interface, we flashed the OpenThread NCP firmware to the same nRF52840 dongle. We implemented the Nety option and added it to the CoAP library of the OpenThread.

Table 1. IoT device specifications

Item	Contents
Module	nRF52840
CPU	32 bit ARM Cortex-M4F (64 MHz)
Flash	1 MB
RAM	256 KB
Network	IEEE 802.15.4 radio

Fig. 7. Devices used for evaluation

Each device has the OpenThread stack and the modified CoAP library. The firmware file server is implemented on the same Raspberry Pi where the border router is running. The server is transmitting CoAP block-wise with the Nety option to the IoT devices.

PLR in a Single Hop Scenario. Unlike the CON type-based unicast communication that the next block can be sent only when the ACK message arrives at the client, a client does not wait for the ACK in multicast communication. Therefore the multicast client can decide how fast it transmits blocks. We show here how packet transmission interval affects reliability in multicast communication over the IEEE 802.15.4 network when there are different number of devices deployed. We set intervals every 5 ms from 30 till 50 ms and tested each case of number of devices 3, 6 and 9. For the evaluation, we transmitted 32,000 blocks for a 200 kB file in 64 Byte size blocks. All the devices are located within 1 hop from the sender. In Table 2, the # of NACK row includes total number of NACK transmitted during the transmission time from all devices and Time row indicates total time for completing the communication.

Table 2. Total transmission time and PLR comparison based on different number of devices and interval

Interval	Test items	Number of devices		
		3	6	9
30 ms	Time	1 min 48 s	1 min 56 s	2 min 8 s
	# of NACK	393	641	782
	PLR	12.28	20.03	24.43
35 ms	Time	1 min 59 s	2 min 14 s	2 min 26 s
	# of NACK	315	582	633
	PLR	9.84	18.18	19.78
40 ms	Time	2 min 20 s	2 min 27 s	2 min 32 s
	# of NACK	294	450	573
	PLR	9.18	14.06	17.90
45 ms	Time	2 min 32 s	2 min 48 s	2 min 54 s
	# of NACK	168	360	413
	PLR	5.25	11.25	12.9
50 ms	Time	2 min 48 s	2 min 53 s	2 min 57 s
	# of NACK	154	244	333
	PLR	4.81	7.62	10.40

In a single hop, the average number of the lost packet per device was not different in each case. However with the same packet loss rate, total amount of NACK messages are increased according to the number of devices. It resulted in increasing time to complete the communication. For the comparison to unicast, we repeated the same file transmission 10 times. In unicast case, the file transmission took 1 min 15 s on average. When there are more than two devices even in 1 hop with 30 ms interval transmission, the firmware file update can be done faster in multicast than unicast reliable way.

PLR in a Multi Hop Scenario. To evaluate the feasibility of our proposal and compare it to unicast in a realistic environment, we deployed 15 nodes in an office building on the same floor and made a mesh network with multi-hops.

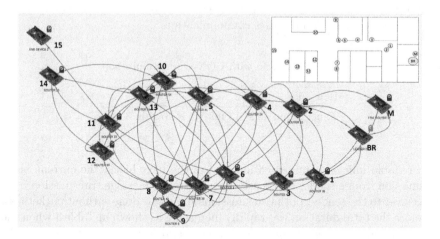

Fig. 8. Device deployment and mesh network topology

Figure 8 shows how the devices made mesh network with other devices. The figure also shows the location of each device on the floor plan. To scan the network topology we used nRF Thread topology monitor tool of Nordic. M is the device used for topology monitoring, and it is not involved in the evaluation. In the mesh network having at most 3 hops from the sender to a receiver, the interval time had to be increased compared to the previous single hop test. To find the best interval time for the topology, we evaluated the PLR with different intervals. To reduce network congestion, when the NACK message arrives from receivers, the server re-transmits the requested packet following the interval.

Table 3. Nety-based multicast in multi hop

Interval	200 ms	250 ms	300 ms
Time	17 min 34 s	14 min 2 s	16 min 8 s
PLR	3.93%	<0.01%	<0.01%

Table 3 indicates that with 200 ms interval, the PLR increased and many packets had to re-transmitted. Thus, even though the interval is shorter, the complete time for the whole communication was longer than the other cases having intervals longer than 200 ms. Furthermore, in case of the interval faster than 200 ms, many devices became bricked since the fast message transmission worked as the Danial of Service (DoS) in the constrained network.

To compare required time to complete the data transmission with unicast case in the same deployment, we evaluated total time for each device located in a different place with different hop number. In our testbed, 5 devices in 1 hop, 9 devices in 2 hops and 1 device in 3 hops are deployed. As a result, it takes in total 2 h 47 min to update all devices with unicast which is 12 times slower than Nety-based reliable multicast communication.

Table 4. Unicast with CON-type in multi hop

Hops	1	2	3
Time	1 min 15 s	13 min 41 s	37 min 30 s

In reliable unicast communication, to get the next block, the current block transmission from a sender to a receiver and an ACK message transmission from the receiver to the sender, both transmissions should be done without packet loss. Therefore the total duration was rapidly increasing as shown in Table 4 when the number of hops is increasing in between sender and receiver. On the other hand, in case of Nety-based multicast communication, as long as one-way transmission which is from the sender to the receiver is successfully done without packet loss, the next block can be transmitted. Thus, especially in a multihop environment, Nety-based multicast communication is far more efficient than unicast-based transmission for big chuck of data transmission.

6 Conclusion

In this paper we proposed a reliable multicast scheme based on a NACK-mechanism for efficient firmware update of constrained IoT devices using a new Nety option that can be easily integrated into CoAP. We demonstrated feasibility by implementing our scheme on real devices and performed an experimental evaluation. We believe our solution can be applied to many multicast IoT use cases with a large number of devices and constrained environments. Adding security to our Nety-based reliable multicast solution remains for further research.

References

1. Adamson, B., Bormann, C., Handley, M., Macker, J.: NACK-oriented reliable multicast (NORM) transport protocol. RFC 5740, RFC Editor, November 2009
2. Bormann, C., Shelby, Z.: Block-wise transfers in the constrained application protocol (CoAP). RFC 7959, RFC Editor, August 2016
3. Chandra, H., Anggadjaja, E., Wijaya, P.S., Gunawan, E.: Internet of Things: over-the-air (OTA) firmware update in lightweight mesh network protocol for smart urban development. In: 2016 22nd Asia-Pacific Conference on Communications (APCC), pp. 115–118. IEEE (2016)

4. Choi, G., Kim, D., Yeom, I.: Efficient streaming over CoAP. In: 2016 International Conference on Information Networking (ICOIN), pp. 476–478. IEEE (2016)
5. Hui, J., Kelsey, R.: Multicast protocol for low-power and lossy networks (MPL). RFC 7731, RFC Editor, February 2016
6. Kauer, F., Meyer, F., Turau, V.: A holistic solution for reliable over-the-air software updates in large industrial plants. In: 2017 13th Workshop on Intelligent Solutions in Embedded Systems (WISES), pp. 29–34. IEEE (2017)
7. Montenegro, G., Kushalnagar, N., Hui, J., Culler, D.: Transmission of IPv6 packets over IEEE 802.15.4 networks. RFC 4944, RFC Editor, September 2007. http://www.rfc-editor.org/rfc/rfc4944.txt
8. Nguyen, J., Yu, W., Ku, D.: Reliable transport for mobile ad hoc networks with constrained application protocol (CoAP) over negative-acknowledgment oriented reliable multicast (NORM). In: 2018 International Conference on Computing, Networking and Communications (ICNC), pp. 361–366. IEEE (2018)
9. Rahman, A., Dijk, E.: Group communication for the constrained application protocol (CoAP). RFC 7390, RFC Editor, October 2014. http://www.rfc-editor.org/rfc/rfc7390.txt
10. Schütz, B., Aschenbruck, N.: Adding a network coding extension to CoAP for large resource transfer. In: 2016 IEEE 41st Conference on Local Computer Networks (LCN), pp. 715–722. IEEE (2016)
11. Shelby, Z., Hartke, K., Bormann, C.: The constrained application protocol (CoAP). RFC 7252, RFC Editor, June 2014. http://www.rfc-editor.org/rfc/rfc7252.txt

A WiVi Based IoT Framework for Detection of Human Trafficking Victims Kept in Hideouts

Sidharth Samanta[1], Sunil Samanta Singhar[2], A. H. Gandomi[3($\boxtimes$)],
Somula Ramasubbareddy[4], and S. Sankar[5]

[1] Department of Computer Science and Applications, Utkal University,
Bhubaneswar, Odisha, India
samantasidharth@gmail.com
[2] Department of Computer Science and Engineering, IIIT Bhubaneswar,
Bhubaneswar, Odisha, India
c119004@iiit-bh.ac.in
[3] Faculty of Engineering and Information Technology, University of Technology
Sydney, Ultimo, Australia
gandomi@uts.edu.au
[4] Department of Information Technology, VNRVJIET, Hyderabad 500090,
Telangana, India
svramasubbareddy1219@gmail.com
[5] Department of Computer Science Engineering, Sona College of Technology,
Salem 636005, Tamilnadu, India
sankarcsharp@gmail.com

Abstract. Human trafficking is the trade of humans for the purpose of forced labor, sexual slavery, or commercial sexual exploitation for the trafficker or others. The traffickers often trick, defraud, or physically force victims into selling sex and forced labor. In others, victims are lied to, assaulted, threatened, or manipulated into working under inhumane, illegal, or otherwise unacceptable conditions. According to the estimation of the International Labor Organization, there are more than 40.3 million victims of human trafficking globally. It is a threat to the Nation as well as to humanity. There have been many efforts by government agencies & NGOs to stop human trafficking and rescuing victims, but the traffickers are getting smarter day by day. From multiple sources, it is observed that the traffickers generally hide humans in hidden rooms, sealed containers, and boxes disguised as goods. This congestion results in Critical mental and physical damages in some cases. It is practically impossible to physically go and check each box, containers or rooms. So in this paper, we propose an Wireless Vision based IoT framework, which uses the reflection of WiFi radio waves generated by WiFi to detect the presence of humans inside a cement or metal enclosure from outside.

Keywords: Human trafficking · WiVi · Object detection · IoT ·
WiFi · Radio frequency · See-through wall · Human presence detection

© Springer Nature Switzerland AG 2020
W. Song et al. (Eds.): ICIOT 2020, LNCS 12405, pp. 96–107, 2020.
https://doi.org/10.1007/978-3-030-59615-6_7

1 Introduction

Nowadays computers, not only outperforming humans in intelligence but also doing such tasks which is impossible for humans. Even conventional things like home appliances, machines, accessories are becoming smarter with the integration of IoT. With every passing day Internet of Things and other technologies like AI, Cloud Computing, Blockchain are setting new milestones in different fields such as health care, education, manufacturing, etc. Apart from these current technologies are also helping law enforcement agencies to combat crime and other unlawful activities. This paper addressed the crime of human trafficking and a noble way to rescue the victims.

1.1 Background

Human trafficking is a trade of human beings for the purpose of slavery, prostitution, and forced labor. It is a global problem affecting people of all ages, victims of sex trafficking in India are predominantly young and illiterate girls average 10–15 years of age, from impoverished families in rural states. It is not only based on sexual or forced labor, but there is a significant rise in the illegal sale of women for purpose of marriage which has become more prevalent in South Asian countries like Pakistan, Bangladesh, and India.

According to the report [1], there are about 20 to 40 million people under slavery and with each passing year, it is increasing by approximately 1 million people. Another report [2], among all victims 71% are female and rest 29% are male. An estimation of 1.2 million children are trafficked each year into exploitative work according to the International Labor Organization. It is estimated that an annual profit of nearly 150 billion USD is earned from human trafficking per as the report [3], from which 99 billion USD comes from prostitution and other sexual exploitation.

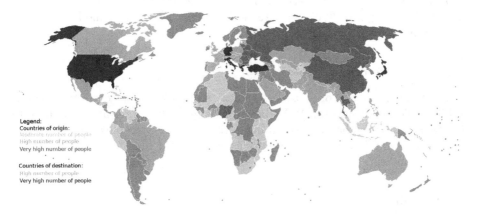

Fig. 1. Origin and destination country schema of global human trafficking [4].

Human trafficking is the most inhuman crime in this world, and its roots are spread across the globe (Fig. 1). It is a type of crime that provides liveware to other crimes, which means it is the initial phase of other crimes such as prostitution, contract killing, terrorism, smuggling, etc. Any action with the human trafficker's network can have a cascading effect on these crimes.

Although the process of trafficking is too complex and varies with time and individual, it can be divided into 4 stages given as follows:

1. Luring
2. Gaming and Grooming
3. Coercion and Manipulation
4. Exploitation

1.2   Problem Statement

During the manipulation and exploitation phase, the traffickers are most venerable in front of law enforcement agencies. To deal with the circumstances, they force the victims to keep on traveling or stay in hideouts. Hideouts are often very small and congested secret rooms, designed to stay hidden during raids. A piece of simple information like the location of a secret place in a house can save millions of lives, but it is pretty impossible to notice each and every hideout during raids. According to the report [5] only 0.04% of all victims have survived or rescued. There are multiple ways to strike down the human trafficking process, but this paper addressed the way of detecting and rescuing the victim.

1.3   Contribution of Research

The literature proposes a WiVi framework build on the top of an IoT device, that can be used to detect human presence. This can be possible because of the reflection and refraction properties of the radio waves. This paper further implemented the state of the art WiVi [6] and WiZ [7] to detect humans through the wall by using SSD [8] object detection model. The proposed model can also be used in different military and nonmilitary operations.

1.4   Structure of the Paper

The rest of the paper is organized as follows. Section 2 elaborates on the theory of the Radio Frequency and WiVi as well as some critical Literature. Section 3 and Sect. 4 provide the proposed framework and the methodology respectively. The approach and future scope are discussed in Sect. 5 and finally the paper is concluded in Sect. 6.

2   Wireless Vision

Popularly known as WiVi [6], it follows the same principle as normal photographic sensors to capture an image, but it uses radio-wave instead of visible light waves. WiVi is compact, low cost, low bandwidth, and easily accessible non-military technology used to sense objects present behind a fully opaque surface. The characteristics of the radio-wave help WiVi to possess such features. Later in this section, the working of WiVi is discussed.

2.1 Radio Wave

Radio wave falls under a frequency range of about 20 kHz to approximately 300 GHz, is the frequency rate of oscillation of an electrical or voltage alternating electric current or magnetic, electromagnetic, or electromagnetic field or mechanical device. This is approximately between the upper limit of audio frequencies and the lower limit of infrared frequencies. These are the frequencies at which oscillating current energy can radiate as radio waves of a conductor into space. Different sources for the frequency spectrum (Fig. 2) define specific upper and lower limits.

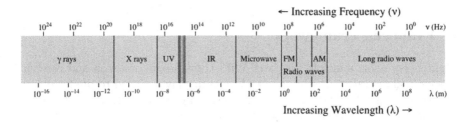

Fig. 2. The electromagnetic spectrum.

Frequency is expressed in units called hertz, which are the cycles per second for the propagation of a wave. The human eyes can not see the frequencies that are beyond the visual spectrum 405–790 THz, but it can be sensed by using specific kinds of sensors. The visual and infrared waves can not penetrate a fully opaque barrier as they interact with the pigment and the temperature of the surface respectively. Except these two, other waves such as radio wave, microwave, x-ray, and gamma-ray can penetrate this kind of surfaces.

2.2 WiVi

WiVi is a device that is used to captures objects present behind a fully opaque surface like a wall. Because of the presence of Wi-Fi chipsets, the through-wall imaging is comparatively cost less, consumes less power, and available to average users. It uses WiFi OFDM signals in 2.4 GHz ISM band and conventional WiFi hardware. It consists of 3 antennas in such a way that two of them are used as transmitter and one as the receiver. The directional antennas focus the radio wave toward the area.

2.3   Literature

Decades before the introduction of WiVi, the curiosity to see-through-wall was introduced [9]. With time the problem is simulations [10,11] and modeled [12,13]. Literature like [14–16] implemented it on moving object detection. After its inception in [6], the legacy continued to provide some critical literature addressing emotion recognition [17], object tracking [18], detection sleep stages [19], motion tracking [20], human pose, skeleton, action and mesh detection [21–24] by the original authors. The work is also deployed in various areas such as surveillance, healthcare, Human-Computer-Interaction, etc.

A violence detection system is proposed in [25] by using WiVi, whereas literature [26,27] proposed to use the wireless technique for crowd counting and occupancy monitoring. A WiVi based person identification framework is proposed in [28] and the literature [29] suggested a WiVi surveillance system based on a drone. In healthcare, wireless vision can help monitoring Heartbeat [30,31] and sleep [32]. According to the literature [33–35] WiVi can become a potential tool for gesture recognition system.

3   Proposed Framework

A microprocessor, multiple arrays of WiFi transmitting antenna and receiving sensors, and actuators like buzzers and LCD display are used to implement the framework (Fig. 3). The transmitting sensors are used to throw radio waves, and the receiving sensors are used to pick up the radio wave reflect back by hitting nearby objects. The sensing data collected by the receiver processed by a microprocessor for human detection and the buzzer is triggered when detected. Specifications and characteristics of these devices are described in the following subsections.

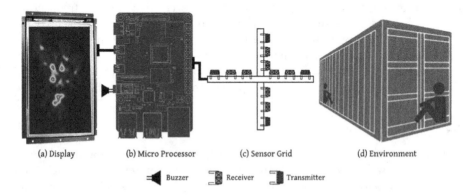

(a) Display   (b) Micro Processor   (c) Sensor Grid   (d) Environment

Buzzer   Receiver   Transmitter

Fig. 3. Physical architecture of the proposed framework.

3.1   Microprocessor

A single-board computer having a quad-core 1.5 GHz microprocessor named Raspberry Pi 4 is used to perform the computation. The task includes controlling the sensor grid, reconstructing the image from sensed data, and detecting objects in real-time. These processes are further elaborated in Sect. 4. This can also be used to transmit the sensor data to a remote server, for precise analysis.

3.2   Antennas and Sensors

The model uses the WiFi transmitting antenna to relay the radio wave and RF sensors for receiving the reflections. As discussed in Sect. 2, the radio wave can move across solid structures like brick cement walls and metal covered containers. These sensors are arranged in a two array grid (Fig. 3c), where the vertically placed sensors collect the perpendicular reflection, and the horizontal sensor collects the parallel radio wave reflection. The framework used multiple transmitting antennas and multiple sensors to gather more angular reflections, which helps in image reconstruction and object localization (Fig. 5).

3.3   Buzzer and Actuators

Though the proposed model uses an LCD display to stream the heat-map video obtained by the sensors and a buzzer to indicate the presence of humans (Fig. 3a). There are many possibilities to add a number of actuators for desired outputs and alerting systems.

4   Method

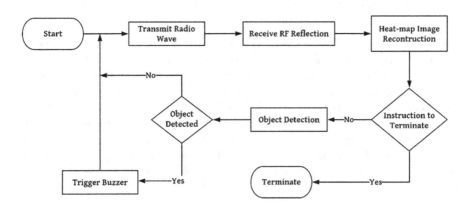

Fig. 4. Workflow of proposed framework.

This section discusses the workflow (Fig. 4) and the method behind the framework proposed in Sect. 3. After the array of RF sensors capture the reflected radio waves, the sensed data further reconstructed as a heat-map image frame. Then the image is processed for noise removal and object detection. For a better understanding of the proposed framework, its workflow is divided into 3 phases, as follows:

1. Signal Acquisition
2. Image Reconstruction
3. Object Detection

4.1   Signal Acquisition

Propagation of RF Signal. Radio wave propagation is the behavior of radio waves as they migrate or propagate from one place to another, or in various areas of the atmosphere. As a form of electromagnetic radiation, such as light waves, radio waves are affected by the phenomena of reflection, refraction, diffraction, absorption, polarization and dispersion. Understanding the impact of changing conditions on radio transmission has many practical applications according to the need.

All electromagnetic (radio, light, X-rays, etc.) waves, in free space, obey the inverse-square rule which specifies that the power density ρ is proportional to that of the square distance r from the point source, or:

$$\rho \propto \frac{1}{r^2}.$$

The transmitter antenna may typically be approximated by a point source at normal contact distances from a transmitter. Doubling the distance of the receiver from the transmitter ensures that the amplitude of the radiated wave at the new position is reduced to one-quarter of its previous size.

Reception Due to Reflection. When a radio wave, or indeed any electromagnetic wave, encounters a change in the medium, some or all of it may spread to the new medium, and the remainder may be reflected. The part that enters the new medium is called the transmitted wave, while the other part is called the reflected wave. In real transmission paths, radio waves are often reflected by a variety of different surfaces. So here the challenging task is to determine the reflected wave received comes from which surface if there is a multiple layer of surfaces. Depending on the interval time of transmission & reception, the energy content in the reflected wave and the texture of the reflected wave it can be determined from which surface it got reflected.

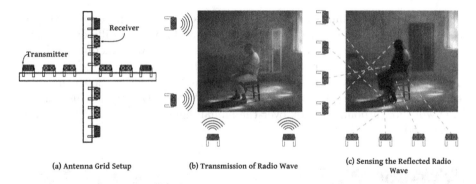

(a) Antenna Grid Setup   (b) Transmission of Radio Wave   (c) Sensing the Reflected Radio Wave

Fig. 5. The process of signal acquisition.

4.2 Image Reconstruction

This sub-phase includes data to heat-map translation and image reconstruction. Heat-maps are great for visualizing statistical and senor data, and it is very easy to produce as well. In hit-map, the warmer color represents a relatively higher value than the cooler color. In the proposed model data collected from sensors translated into heat-map. As the model used unidirectional sensors for accuracy, as it collects more accurate and precise data from its line of sight (Fig. 6a and b). Then these heat-map data are merged together to form a single image frame (Fig. 6c).

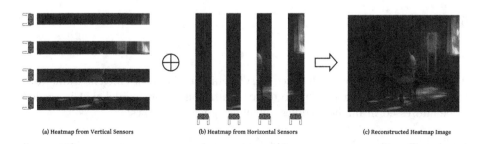

(a) Heatmap from Vertical Sensors   (b) Heatmap from Horizontal Sensors   (c) Reconstructed Heatmap Image

Fig. 6. The process of heat-map image reconstruction.

4.3 Object Detection

This sub-phase includes training dataset generation and object detection (Fig. 7). A specially trained MobileNetV3-SSD object detection framework build with TensorFlow is deployed on the Raspberry pi. Unlike YOLOs and RCNNs, SDD is not resource-hungry. Here the accuracy of the object detection model is traded off due to the limitation of processing power. The model is trained with a heat-map image data set, which is created by annotating human figures in a large number of heat-map images, which are captured by the proposed device itself in a different environment having different human forms and different kinds of barriers between humans and sensors like a brick cement wall, metal containers.

Although SSD is a less resource-consuming object detection model, real-time implementation can put a lot of strain on CPU and overheating. So the model is set to process the video at or less than 8 FPS. When SSD detects a human, the LCD will show a bounding box over the human figure (Fig. 7c) and the buzzer is triggered.

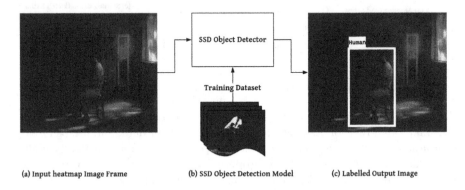

(a) Input heatmap Image Frame   (b) SSD Object Detection Model   (c) Labelled Output Image

Fig. 7. The process of object detection.

5   Discussion

5.1   Limitations

Although the proposed model used WiVi, which is the most efficient framework to see through wall, there are two major limitations of our model that is needed to be addressed in further studies. They are given as follows:

Radio-Wave Noise. When WiFi antenna throws the radio wave, Each and every object of the environment reflect distinct amount of wave back to the sensor. Any reflection, other than the reflection of the desired object is a noise. The noise level increases, when the environment also contain a WiFi.

Limited Resources. As the model is built on a mobile single board computer, so resources like processing power, memory, and available energy are limited. Connecting to the cloud will result in precise detection, but the network latency will preempt its real-time status.

5.2 Future Scope

There are numerous ways, how the proposed framework can be extended or deployed. Unlike a mobile device as proposed in the paper, the framework can be implemented as a fixed devise, placed in a toll booth to scan vehicles for monitoring vehicle occupancy and cargo contents. The device can also be used in many applications, such as the detection of electronic devices in examination hall, or intruder alerting system.

6 Conclusion

Human trafficking is a serious crime and needs to be addressed as strongly as possible. It is threat to not only to the society, but to humanity as well. It is the most profitable crime spread across globe and can be denoted as mother of all crimes as it provides workforce to other crimes such as prostitution, organ stealing, drug paddling, illegal migration, child exploitation, contract killing etc. That means reducing the footprint of human trafficking can affect other criminal activities too. Locating and rescuing victims is a very difficult process, and with each passing days the traffickers are becoming more smart.

Since the inception of technology, it is helping to solve the human task more efficiently. The framework proposed in this paper is an effective way to detect and rescue human trafficking victims. It uses the harmless, low cost and most efficient wireless technology available for non-military entities. There are several limitations in the model, which needed to be addressed in further studies.

References

1. Forced Labor, Modern Slavery, and Human Trafficking. International Labor Organization. http://www.ilo.org/global/topics/forced-labour/lang-en/index. html. Accessed 1 May 2020. Monitoring Target 16.2 of the United Nations Sustainable Development Goals. United Nations Office on Drug and Crime. https://www.unodc.org/documents/research/UNODC-DNR_research_brief.pdf. Accessed 1 May 2020
2. What is Human Trafficking. Californians Against Sexual Exploitation. http:// www.caseact.org/learn/humantrafficking/. Accessed 1 May 2020
3. Human Trafficking by the Numbers. Human Rights First. https://www. humanrightsfirst.org/resource/human-trafficking-numbers. Accessed 1 May 2020
4. Human trafficking. https://en.wikipedia.org/wiki/Human_trafficking. Accessed 1 May 2020
5. Trafficking and Slavery Fact Sheet. Free the Slaves. https://www.freetheslaves. net/wp-content/uploads/2018/04/Trafficking-ans-Slavery-Fact-Sheet-April-2018. pdf. Accessed 1 May 2020
6. Adib, F., Katabi, D.: See through walls with WiFi!. In: Proceedings of the ACM SIGCOMM 2013 Conference on SIGCOMM, 27 August 2013, pp. 75–86 (2013)
7. Adib, F., Kabelac, Z., Katabi, D.: Multi-person motion tracking via RF body reflections (2014)

8. Liu, W., et al.: SSD: single shot multibox detector. In: Leibe, B., Matas, J., Sebe, N., Welling, M. (eds.) ECCV 2016. LNCS, vol. 9905, pp. 21–37. Springer, Cham (2016). https://doi.org/10.1007/978-3-319-46448-0_2

9. RadarVision. http://www.timedomain.com.TimeDomainCorporation

10. Wang, H., Narayanan, R., Zhou, Z.: Through-wall imaging of moving targets using UWB random noise radar. IEEE Antennas Wirel. Propag. Lett. **8**, 802–805 (2009)

11. Ram, S., Li, Y., Lin, A., Ling, H.: Doppler-based detection and tracking of humans in indoor environments. J. Franklin Inst. **345**, 679–699 (2008)

12. Soldovieri, F., Solimene, R.: Through-wall imaging via a linear inverse scattering algorithm. IEEE Geosci. Remote Sens. Lett. **4**, 513–517 (2007)

13. Solimene, R., Soldovieri, F., Prisco, G., Pierri, R.: Three-dimensional through-wall imaging under ambiguous wall parameters. IEEE Trans. Geosci. Remote Sens. **47**, 1310–1317 (2009)

14. Ralston, T., Charvat, G., Peabody, J.: Real-time through-wall imaging using an ultra wide band multiple-input multiple-output (MIMO) phased array radar system. In: IEEEARRAY (2010)

15. Yang, Y., Fathy, A.: Design and implementation of a low-cost real-time ultra-wide band see-through-wall imaging radar system. In: IEEE/MTT-S International Microwave Symposium (2007)

16. Chetty, K., Smith, G., Woodbridge, K.: Through-the-wall sensing of personnel using passive bi-static WiFi radar at stand off distances. IEEE Trans. Geosci. Remote Sens. **50**, 1218–1226 (2012)

17. Zhao, M., Adib, F., Katabi, D.: Emotion recognition using wireless signals. In: Proceedings of the 22nd Annual International Conference on Mobile Computing and Networking, 3 October 2016, pp. 95–108 (2016)

18. Adib, F., Kabelac, Z.E., Katabi, D., inventors; Massachusetts Institute of Technology, assignee: Object tracking via radio reflections. United States patent application US 15/120,864, 16 March 2017

19. Zhao, M., Yue, S., Katabi, D., Jaakkola, T.S., Bianchi, M.T.: Learning sleep stages from radio signals: a conditional adversarial architecture. In: Proceedings of the 34th International Conference on Machine Learning, 6 August 2017, vol. 70, pp. 4100–4109 (2017). JMLR.org

20. Adib, F., Kabelac, Z.E., Katabi, D., inventors; Massachusetts Institute of Technology, assignee: Motion tracking via body radio reflections. United States patent US 9,753,131, 5 September 2017

21. Zhao M., et al.: Through-wall human pose estimation using radio signals. In: Proceedings of the IEEE Conference on Computer Vision and Pattern Recognition, pp. 7356–7365 (2018)

22. Zhao, M., et al.: RF-based 3D skeletons. In: Proceedings of the 2018 Conference of the ACM Special Interest Group on Data Communication, 7 August 2018, pp. 267–281 (2018)

23. Li, T., Fan, L., Zhao, M., Liu, Y., Katabi, D.: Making the invisible visible: action recognition through walls and occlusions. In: Proceedings of the IEEE International Conference on Computer Vision, pp. 872–881 (2019)

24. Zhao, M., et al.: Through-wall human mesh recovery using radio signals. In: Proceedings of the IEEE International Conference on Computer Vision, pp. 10113–10122 (2019)

25. Zhang, L., Ruan, X., Wang, J.: WiVi: a ubiquitous violence detection system with commercial WiFi devices. IEEE Access **8**, 6662–6672 (2019)

26. Xi, W., et al.: Electronic frog eye: counting crowd using WiFi. In: IEEE INFOCOM 2014-IEEE Conference on Computer Communications, 27 April 2014, pp. 361–369. IEEE (2014)
27. Depatla, S., Muralidharan, A., Mostofi, Y.: Occupancy estimation using only WiFi power measurements. IEEE J. Sel. Areas Commun. **33**(7), 1381–1393 (2015)
28. Zeng, Y., Pathak, P.H., Mohapatra, P.: WiWho: WiFi-based person identification in smart spaces. In: 2016 15th ACM/IEEE International Conference on Information Processing in Sensor Networks (IPSN), 11 April 2016, pp. 1–12. IEEE (2016)
29. Sehrawat, A., Choudhury, T.A., Raj, G.: Surveillance drone for disaster management and military security. In: 2017 International Conference on Computing, Communication and Automation (ICCCA), 5 May 2017, pp. 470–475. IEEE (2017)
30. Saha, H.N., Nandi, M., Biswas, U., Das, T.: Heart-rate detection and tracking human body movements through walls at home. In: 2016 IEEE 7th Annual Information Technology, Electronics and Mobile Communication Conference (IEMCON), 13 October 2016, pp. 1–4. IEEE (2016)
31. Khan, F., Bhuiyan, M.Z., Islam, M.M., Wang, T., Zaman, A., Tao, H.: Wi-Fi signal analysis for heartbeat and metal detection: a comparative study of reliable contactless systems. In: 2019 6th International Conference on Dependable Systems and Their Applications (DSA), 3 January 2020, pp. 81–90. IEEE (2020)
32. Liu, X., Cao, J., Tang, S., Wen, J.: Wi-Sleep: contactless sleep monitoring via WiFi signals. In: 2014 IEEE Real-Time Systems Symposium, 2 December 2014, pp. 346–355. IEEE (2014)
33. Li, H., Yang, W., Wang, J., Xu, Y., Huang, L.: WiFinger: talk to your smart devices with finger-grained gesture. In: Proceedings of the 2016 ACM International Joint Conference on Pervasive and Ubiquitous Computing, 12 September 2016, pp. 250–261 (2016)
34. He, W., Wu, K., Zou, Y., Ming, Z.: WiG: WiFi-based gesture recognition system. In: 2015 24th International Conference on Computer Communication and Networks (ICCCN), 3 August 2015, pp. 1–7. IEEE (2015)
35. Wang, J., Vasisht, D., Katabi, D.: RF-IDraw: virtual touch screen in the air using RF signals. ACM SIGCOMM Comput. Commun. Rev. **44**(4), 235–246 (2014)

Spatio-Temporal Coverage Enhancement in Drive-By Sensing Through Utility-Aware Mobile Agent Selection

Navid Hashemi Tonekaboni[1]([✉]), Lakshmish Ramaswamy[2], Deepak Mishra[2], Omid Setayeshfar[2], and Sorush Omidvar[2]

[1] College of Charleston, Charleston, SC 29424, USA
hashemin@cofc.edu
[2] University of Georgia, Athens, GA 30602, USA
{laksmr,dmishra,omid.s,omidvar}@uga.edu

Abstract. In recent years, the drive-by sensing paradigm has become increasingly popular for cost-effective monitoring of urban areas. Drive-by sensing is a form of crowdsensing wherein sensor-equipped vehicles (aka, mobile agents) are the primary data gathering agents. Enhancing the efficacy of drive-by sensing poses many challenges, an important one of which is to select non-dedicated mobile agents on which a limited number of sensors are to be mounted. This problem, which we refer to as the mobile-agent selection problem, has a significant impact on the spatio-temporal coverage of the drive-by sensing platforms and the resultant datasets. The challenge here is to achieve maximum spatiotemporal coverage while taking the relative importance levels of geographical areas into account. In this paper, we address this problem in the context of the SCOUTS project [1], the goal of which is to map and analyze the urban heat island phenomenon accurately.

Our work makes several significant technical contributions. First, we delineate a model for representing the mobile agent selection problem. This model takes into account the trajectories of the vehicles (public transportation buses in our case) and the relative importance of the urban regions, and formulates it as an optimization problem. Second, we provide an algorithm based on the utility (coverage) values of mobile agents, namely, a hotspot-based algorithm that limits the search space to important sub-regions. Third, we design a highly efficient coverage redundancy minimization algorithm that, at each step, chooses the mobile agent, which provides maximal improvement to the spatio-temporal coverage. This paper reports a series of experiments on a real-world dataset from Athens, GA, USA, to demonstrate the effectiveness of the proposed approaches.

Keywords: Spatiotemporal data analysis · Drive-by sensing · Coverage enhancement

W. Song et al. (Eds.): ICIOT 2020, LNCS 12405, pp. 108–124, 2020.
https://doi.org/10.1007/978-3-030-59615-6_8

1   Introduction

The massive proliferation of mobile sensor devices is changing the landscape of environmental monitoring by augmenting conventional data sources such as satellites and weather stations with the crowdsensing paradigm. In particular, crowdsensing is very beneficial in urban areas where higher population densities not only provide larger pools of potential contributors but also enhances the impact of crowdsensing on the local population.

Although people are considered to be the main participants in crowdsensing, a new category of this paradigm, namely *drive-by sensing*, has recently emerged. In the drive-by sensing paradigm, the primary sensing agents are vehicle-borne sensors [2]. Drive-by sensing has numerous applications in urban and environmental monitoring, especially where the properties that are being monitored exhibit strong spatio-temporal associations. Google street view [3] is a famous example of the drive-by sensing paradigm, wherein vehicles are employed to collect street-level imagery on a global scale. Drive-by sensing paradigm has also been employed to monitor the road conditions at both the surface and subsurface levels [4].

A recent research direction has been to employ non-dedicated vehicles (vehicles whose primary functionality is not sensing/data gathering) for drive-by sensing. Here, sensing occurs *opportunistically* during the regular operation of the vehicles. For example, in the City Scanner project [5], sensors were mounted on municipal garbage trucks to collect a multitude of environmental parameters of the city without interfering with the routes or operations of the truck fleet.

Drive-by sensing through public transportation vehicles (e.g., city buses) is attractive because of the several advantages it offers. First, these vehicles frequently move around the cities throughout the day, providing a cost-effective means to monitor large swathes of cities. Second, because these vehicles ply on pre-defined routes and follow pre-defined schedules, it is possible to estimate their locations at a given time of day. This permits systematically planned data gathering. Since these vehicles return to specific locations at the end of their shifts, scheduling the mounting and maintenance of sensors becomes less cumbersome.

While drive-by sensing through non-dedicated vehicles is becoming popular, making it effective, efficient and practical poses significant difficulties such as lack of control on the routes and the schedules of these vehicles, their uneven spatio-temporal sensing coverage, and the high costs along with the human efforts involved in installing and maintaining the sensors on vehicles. One of the major research challenges is to select a subset of public transportation vehicles (i.e., buses) to mount the sensor devices. This challenge acquires importance because budgetary constraints and human effort required in installing and maintaining sensors often limit the number of sensors that can be deployed (i.e., it is impractical to deploy sensors on all city buses). For instance, a single sensor setup to monitor urban temperatures costs more than a hundred dollars, and installing and configuring a setup on a bus requires a few hours of work from a human expert.

In this context, it is imperative to design a cost-effective strategy to select buses for installing a limited number of sensors so as to maximize the benefits of sensing in terms of spatio-temporal granularity and coverage of the sensed data. In this paper, we refer to this as the *mobile agent selection problem*. Furthermore, based upon the urban phenomena being monitored, certain parts of the city may have a higher importance in the sense that they may need to be sensed at higher spatio-temporal resolutions. For example, these may be densely populated regions or regions with significant variations in the environmental parameters. Thus, the mobile agent selection problem (city buses in the context of this study) has to take into account the relative importance levels of the sub-regions of an urban area.

In this paper, we focus on this problem in the context of the SCOUTS project [1], the goal of which is to generate hyperlocal heatmaps of urban regions with high spatio-temporal granularity. In addressing this problem, we make several novel technical contributions, which can be summarized as follows:

- We provide a novel mathematical formulation for the mobile agent selection problem. This model is unique in the sense that it takes into account the trajectories of vehicles as well as the relative importance levels of various sub-regions of an urban area. We formulate this as a constrained optimization problem with an objective function that encapsulates spatio-temporal sensing granularity requirements.
- We propose our hotspot-based algorithm that shrinks the spatial grid of the whole city to a grid of hotspot cells, which have higher relative importance compared to other grid cells.
- Third, we design a highly efficient redundancy minimization algorithm. At each step, this algorithm chooses the bus that provides maximal improvement to the spatio-temporal coverage of the current selection. This is done by minimizing the redundancies caused by overlapping trajectories. This algorithm not only outperforms the above algorithm in terms of the spatio-temporal sensing coverage but also runs orders of magnitude faster than an exhaustive search approach.

We evaluate the performance of all the proposed algorithms on a real-world bus trajectory dataset from the public transit system from Athens, Georgia, USA (ACC public transit). Our experiments show that our proposed algorithms significantly enhance the spatio-temporal coverage of a limited number of sensing agents.

2   Background and Motivation

2.1   SCOUTS Project

The SCOUTS project at the University of Georgia aims to generate hyperlocal heat exposure maps of different urban communities [1]. For this purpose, we augment traditional data sources such as satellites and weather stations with human-borne crowdsensing and drive-by sensing. Public transportation buses were deemed to the most suitable agents for drive-by sensing because of two reasons. On the one hand, the routes of these buses are close to the daily commute of city-dwellers. On the other hand, the constant movement of these vehicles throughout the cities provide reasonable spatio-temporal coverage of urban regions.

Figure 1 shows the picture of a temperature sensor setup mounted on public transportation buses in Athens, GA. The sensor setup was assembled in-house by our project team. It consists of an Arduino microcontroller board, DS18B20 1-wire digital temperature sensors with 0.5C accuracy, a low-power GPS sensor, and lithium-ion batteries in a shielded setting. The cost of each sensor setup is approximately US$120. Apart from the maintenance costs, assembling and mounting each sensor setup requires approximately four man-hours.

Fig. 1. Temperature sensors mounted on city buses

2.2 Mobile Agent Selection Problem

As stated in the introduction, budgetary and human resource constraints often limit the number of sensor kits that can be mounted in a city's transportation system. On the other hand, the subset of buses that are selected for carrying the sensors has a significant impact on the spatio-temporal coverage of the drive-by sensing platform.

In order to demonstrate the importance of the bus selection platform, let us consider the ACC public transit system. This system consists of 20 city buses covering the Athens city area ($310\,\text{km}^2$), and the whole region is modeled as a rectangular grid with 90 by 90 m square cells. Let us consider a 5-h time window between 9:00 AM and 2:00 PM on 10-02-2018 (a typical weekday).

Figure 2 shows the spatio-temporal sensing coverages of the *best* and *worst* possible bus selections when the number of available sensor kits were 3, 4, 5, and 6 respectively (we assumed the number of sensors is limited to 15% to 30% of the number of buses). We define spatio-temporal coverage as follows: if any given grid cell has at least one sensor reading from the region within its boundary (i.e., at least one sensor-carrying bus passed through the cell) in each non-overlapping 1-h time slots within the 5-h time window (10:00 AM to 11:00 AM, 11:00 AM to 12:00 PM and so on), the spatiotemporal coverage of those sensing agents is the maximum. Note that the spatio-temporal coverage has a direct bearing on the quality of the resultant heat map because it represents the completeness of the underlying data (i.e., if the spatio-temporal coverage is low, it implies a higher percentage of missing data values and vice versa).

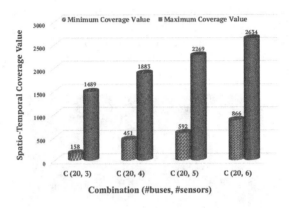

Fig. 2. The worst and the best sensing coverages

Multiple studies have focused on enhancing the sensing coverage in mobile crowdsensing. These studies assume that all the participants are already equipped with the required sensing devices, and they investigate various approaches to distribute sensing tasks while minimizing recruitment costs.

Guo et al. [6] propose a worker selection approach under two situations: either based on the intentional movement of sensing agents for time-sensitive tasks or based on their unintentional movement for tasks which are not time-sensitive. They show how their proposed algorithm outperforms the previous approaches like that of discussed in [7] as a particle swarm optimization solution. In another study, Campioni et al. [8] analyze recruitment algorithms aimed at selecting participants within a crowdsensing network in a way that the most sensing data is obtained for the lowest possible cost. He et al. [9] present a new participant recruitment strategy for drive-by sensing by predicting the future trajectory of participants. Their proposed algorithms show some improvement in terms of crowdsensing coverage.

In another study, Yi et al. [10] propose a fast algorithm for vehicle participant recruitment problem, which achieves a linear-time complexity at the sacrifice of a slightly lower sensing quality. In a separate study, Wang et al. [11] propose a system model based on the predictable trajectory of public transports through a cloud management platform that interacts with static base stations for distributing the sensing tasks. This research, like the other studies discussed in this section, assumes that all the vehicles are equipped with the required sensors and receive a reward per each sensing task.

3   Modeling the Mobile Agent Selection Problem

The overall goal of this model is to maximize spatio-temporal coverage of the data set collected through drive-by sensing while taking the relative importance of hotspot locations into account. Therefore, this model focus on selecting an optimal subset of buses in a way to consider the requirements mentioned above. For this purpose, we assume that the trajectory data of the buses are available. In other terms, the routes that each bus traverse are known. Consequently, by using the GPS data and the timestamps associated with them, we can estimate the location of each bus at a particular point in time.

To formulate the mobile agent selection problem, we model the area as a grid of square cells, as shown in Fig. 3. Each cell is characterized by the GPS coordinate of its four corners. The dimension of each cell is a configurable parameter and represents the spatial granularity of the sensing. We define matrix A, where an arbitrary cell of the grid is represented as a_{ij}:

$$A = \begin{bmatrix} a_{11} & \cdots & a_{1n} \\ : & & : \\ a_{m1} & \cdots & a_{mn} \end{bmatrix}$$

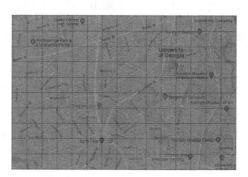

Fig. 3. A sample grid representation

The relative importance of hotspot areas in a city is represented by the weights assigned to their corresponding cells in the grid structure. Therefore, matrix W is defined where each grid cell is associated with a weight:

$$W = \begin{bmatrix} w_{11} & \cdots & w_{1n} \\ : & & : \\ w_{m1} & \cdots & w_{mn} \end{bmatrix}$$

In our design, time is modeled as a vector of $T = \{t_1, t_2, ..., t_l\}$ where each t_k is a time slot with configurable duration. The sum of these time slots is 24 h, and each slot's duration represents the granularity along the temporal dimension. For example, if we need to have a reading of an environmental feature every 30 min, each t_k would denote a 30-min time slot.

The set of $B = \{b_1, b_2, ..., b_p\}$ represents all the buses available in the city where each b_λ represents an individual bus. If a bus b_λ carries a sensor (i.e., it is selected for sensor deployment), it can obtain a reading from the cell a_{ij} in the time slot t_k if and only if b_λ is present within a_{ij}'s boundaries for at least some duration of time slot t_k (i.e., b_λ has traversed through a_{ij} in time slot t_k). Please note that a bus can traverse through multiple cells during a time slot. Also, multiple buses can traverse through the same cell during the same time slot (in which case, we obtain duplicate values). Considering the limited Number of Sensors (NS), we define Bus Set:

$$BS = \{BS_1, BS_2, ..., BS_q\}$$

as the set of all possible bus combinations, where each BS_i is a set of buses ($BS_i \subseteq B$) and the size of each of these sets is less than or equal to the number of available sensors ($|BS_i| \leq NS$). For instance, BS_1 can be represented as:

$$BS_1 = \{b_5, b_{18}, b_{24}\}$$

Objective Function. In this section, we define an objective function for bus selection that reflects the overall goal of maximizing the spatiotemporal coverage. For this purpose, let's suppose that the Selected Bus Set of $SBS^* = \{b_l, b_k, b_p\}$ represents the set of 3 buses which are selected for sensor deployment, such that $SBS \subseteq B$ and $|SBS| \leq NS$. Having laid out the model, we now define the Coverage Value (CV) of BS with respect to a cell a_{ij} at a time slot t_k as follows:

$$CV(BS_x, a_{ij}, t_k) = \begin{cases} w_{ij}^{t_k}, \text{if} \{\exists b_i \in BS_x | b_i \text{ is in } a_{ij} \text{ at } t_k\} \\ 0, \text{otherwise} \end{cases} \tag{1}$$

In other words, CV determines whether at least one of the buses in a set sensed the given cell at the given time slot or not. If the condition is true, the bus set gains the coverage value associated with that location, which is equal to the cell's weight. Otherwise, the set gains no coverage value for that specific time slot.

In the next step towards our objective function, we define the Cumulative Coverage Value (CCV) of a bus set as:

$$CCV(BS_x) = \sum_{t_k \in T} \sum_{\forall a_{ij} \in A} CV(BS_x, a_{ij}, t_k) \tag{2}$$

This measure calculates the aggregated coverage value of a bus set in all the time slots during a day while eliminating the duplicate values. In other terms, if more than one bus in a set covers a grid cell in the same time slot, the weight associated with it will be added to the CCV only once.

Finally, the Selected Bus Set (SBS^*) will be the bus set that its CCV is higher than all other possible bus combinations. If more than one set achieves the same maximum CCV, the set in which its minimum CV in all the time slots is higher than that of the other sets, will be chosen (it denotes the set with better spatial coverage in each single time slot). Therefore, our objective function is defined as follows:

$$SBS^* = BS_x, \text{if } (CCV(BS_x) > \bigwedge_{BS_i - \{BS_x\}} CCV(BS)) \tag{3}$$

In short, the primary motivation is to minimize the redundant values in both space and time by selecting the best subset of our mobile agents.

Illustrative Example. In order to better understand the definitions mentioned above, Fig. 4 shows a sample grid with 16 cells and no hotspot $(w_{ij} = 1)$. The routes that each bus passed during a time slot is depicted using the dotted lines. Let's suppose that there are two bus selections named BS_1 and BS_2, where:

$$BS_1 = \{bus_1, bus_2, bus_3\}$$

$$BS_2 = \{bus_3, bus_4, bus_5\}$$

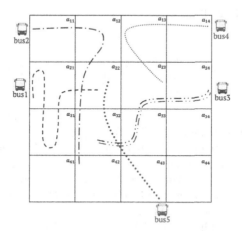

Fig. 4. An example of a bus selection coverage in one time slot

Although bus_3 is selected in both sets, the other two buses are different. Table 1 represents the number of cells passed by each bus during a given time slot. For instance, bus_1 passed three different cells $(a_{21}, a_{31}, \text{and } a_{22})$; therefore, it gets the coverage value of 3. Similarly, the coverage value for other buses is calculated. The last column represents the sum of the coverage values while excluding the duplicates. Therefore, although the actual sum of the values in the first row equals 12, eliminating the duplicates reduces it to 10. As depicted in Fig. 4, we can see that the two cells of a_{22}, a_{32} are covered twice. Thus, the first bus set as a whole, gained a coverage value of 10 in this time period.

In the next step, we want to continue with the same example in Fig. 4 for three consecutive time slots. In Fig. 5 the grid on the back corresponds to the same bus set of BS_1 which we saw in Fig. 4. Considering that during the first time slot, BS_1 covered 10 different cells, this selection gains the coverage values of 10 in t_1. During the second time slot, the three buses continued their routes and sensed 12 different cells. Although some of the cells were already sensed during t_1, these cells are counted again in t_2, because we only exclude the overlaps within the same time slot. Therefore, BS_1 gets the coverage value of 12 in t_2. Following the same logic, BS_1 gains the coverage value of 9 during t_3. These coverage values correspond to the first row of Table 2.

Table 1. Calculating bus coverage value at t_l

w_{ij}	bus_1	bus_2	bus_3	bus_4	bus_5	ΣCV
BS_1	3	5	4	-	-	10
BS_2	-	-	4	3	4	8

Table 2. Total sensed cells per each sensing period

ΣCV	t_1	t_2	t_3
BS_1	10	12	9
BS_2	8	10	10
BS_3	10	11	10

In the next step, we generate the Total Coverage Value for each bus set during the whole time period. The first column in Table 3 represents the CCV for each BS_i. The values of this column are simply the summation of the values in each row of Table 2. The second column of Table 3 shows the minimum value of each row of Table 2. In other terms, this column shows the minimum coverage values that each bus set earned during each time slot.

To better understand how the different weights of hotspot locations can affect the sensing coverage values, Fig. 6 depicts the previous example with BS_1 and BS_2 while the grid cells have different weights. Accordingly, Table 4 shows the updated coverage values of these two bus sets at t_1.

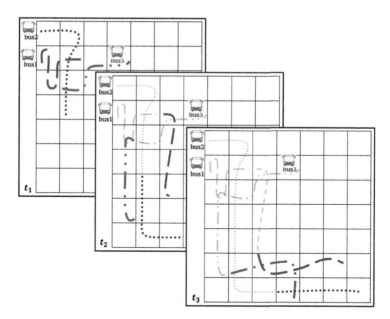

Fig. 5. An example of a bus selection (BS_1) coverage in three consecutive time slots

Table 3. Total sensing coverage value for each bus selection during the whole time period

	CCV(BS_x)	MinCV(BS_x)
BS_1	31	9
BS_2	28	8
BS_3	31	10

Table 4. Calculating bus selection coverage value at t_l with hotspots

w_{ij}	bus_1	bus_2	bus_3	bus_4	bus_5	$\sum$CV
BS_1	9	3	13	-	-	23
BS_2	-	-	13	7	22	30

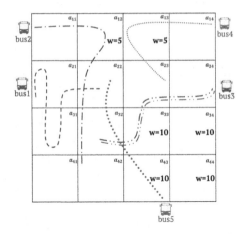

Fig. 6. An example of a grid with hotspots of different weights

4   Simple Approaches and Their Limitations

4.1   Naive Approach

The simplest approach to solve the problem is to mount sensors on a randomly selected set of buses. Since this approach does not consider any requirement with respect to spatiotemporal coverage or the hotspot locations, it will often lead to a poor selection of buses. Furthermore, the variance of CCV among different runs of the algorithm will be very high.

4.2   Exhaustive Approach

The other approach which considers our discussed objective function is the exhaustive method. In this approach, all the possible combinations of n buses taken r at a time, where r is equal to the number of sensors (NS), is computed. Then, the bus combination with the highest CCV will be chosen.

In this algorithm, we first need to create the grid structure based on the given size for each cell by using the latitude and longitude of the area. Next, the algorithm generates the matrix W, where the weights associated with each grid are provided by domain scientists based on the target phenomena to be monitored. Besides, this algorithm creates all the possible bus selections with X different buses where X is equal to or less than the number of sensors. Furthermore, it calculates the set of time slots within the total sensing period. Given these data, the main function chooses the most optimal bus set with the highest spatial and temporal coverage.

This algorithm calls two other functions. The first function, which is called *CCV_Calculation*, determines the cumulative coverage value earned by each given bus set by looping through the set of mobile agents, the cells within the grid structure, and the weights associated with each grid cell. Furthermore, it calculates the minimum coverage value during different time slots for each grid cell. The second function, called SBS, chooses the best selection by applying the objective function. In other terms, it finds the bus selection with the highest CCV.

Although this method is computationally expensive (its runtime grows factorially in terms of the number of bus combinations), it is guaranteed to choose the best possible bus combination where the CCV is higher than all other bus sets.

Considering that the exhaustive approach calculates all the r–combinations of the set of mobile agents where r is the limited number of sensors, running the algorithm for large data sets leads to extremely long processing time. There are many applications where the sensing parameters, such as the coverage values associated with each hotspot, changes quickly. Thus, we have to unmount and mount our sensors on a new subset of buses to monitor the target environmental features in a dynamic setting. For instance, a football game may necessitate extra surveillance coverage in some specific locations around the stadium. Therefore, there should be mechanisms to select an optimal subset of buses to mount surveillance cameras for monitoring the areas of interest for that particular day. As a result, there is a need for utility-aware approaches with a fast decision process to choose an optimal subset of public transportation vehicles to cover the target areas.

To provide a better understanding of the scale of real-world applications, Table 5 provides the number of buses in some selected cities around the world [12–17]. It also represents the number of different bus combinations if 5%, 10%, or 20% of the buses are supposed to be selected. For instance, there are 639 buses in Atlanta. If we want to select 32 buses out of 639 which traverse around this city, we need to calculate the CCV of around 1.03E+54 different bus selections.

Table 5. Combinations of different bus selection in selected cities

City	Number of Buses	Number of Sensors (~5%)	Number of Combinations	Number of Sensors (~10%)	Number of Combinations	Number of Sensors (~20%)	Number of Combinations
Atlanta	639	32	1.03E+54	64	1.1E+89	128	3.8E+137
New York City, Tokyo	1400	70	2.4E+119	140	1.6E+196	280	4.7E+302
Washington	1500	75	9.9E+127	150	2.09E+210	300	2.4E+324
Los Angeles	2400	120	3.1E+205	240	1.9E+337	480	7.6E+519
Karachi	7400	370	2.0E+636	740	8.5E+1042	1480	1.8E+1606
Beijing	24347	1217	4.7E+2096	2435	3.7E+3435	4869	5.1E+5288

5 Utility-Aware Approaches

In this section, we propose our utility-aware sensing approaches to resolve the limitations associated with the simple approaches, which can be leveraged in various sensing frameworks where selecting a subset of vehicles is required.

5.1   Hotspot-Based Approach

The hotspot-based approach is designed based on the relative importance of some areas in an urban region. The importance of a particular area is indicated by a weight (w_{ij}) assigned to the corresponding grid cell, while the default weight of each grid cell is one.

In this approach, instead of running the aforementioned exhaustive algorithm on all the grid cells, we only consider hotspot cells, i.e., the cells that correspond to areas with higher importance levels as indicated by their respective weight values. The threshold of these weight values for a cell to be considered a hotspot is a configuration parameter and is specified at the time of running the algorithm. For instance, the locations with a high variation in temperature or the areas with high population density can be configured as hotspot locations.

On the other hand, this approach excludes the buses that do not pass through any hotspot location. Excluding buses that do not pass through hotspots significantly reduces the number of bus combinations that need to be considered, thus letting the algorithm to perform much faster.

5.2   Utility-Aware Redundancy Minimization Algorithm

Our redundancy minimization algorithm is designed as follows: in the first step, the bus that has the highest spatiotemporal coverage and goes through the largest number of hotspots in different time slots is selected. In the next step, the algorithm chooses the second bus with the best coverage while it passes through the most number of remaining hotspots. In other words, in each step, the selection is made in a way to maximize the number of sensed hotspots, excluding the ones which are already covered by the previous mobile agents. This algorithm continues selecting one mobile agent at a time until it reaches the limit imposed by the number of sensors.

As shown in the pseudocode of our utility-aware spatiotemporal redundancy minimization algorithm in Fig. 7, the algorithm starts by choosing the bus with the highest CCV. Upon selecting a bus, the weight associated with all the grid cells where that specific bus covered in each time slot will be changed to zero. Therefore, by excluding the overlaps, the next bus will be selected such that it covers the highest number of remained cells and hotspots.

In general, greedy strategies, like our redundancy minimization algorithm, are promisingly efficient in analyzing large spatiotemporal data sets [18]. In particular, these approaches become very useful for solving combinatorial optimization problems. Big data analysis necessitates leveraging scalable computational methods that can be used in real-world scenarios where a fast and efficient decision-making process is required.

```
UTILITY-AWARE GREEDY ALGORITHM (A,W,T, BS)

Parameters:
       A – The grid of the area
       W – The matrix of weights based on hotspots
       T – The list of time slots
       B – List of buses
       NS – Number of Sensors
Returns:
       SBS* – The best bus combination

for i=0 till i < NS ; i++
       CCV← CCV_Calculation (A, W, T, B)
       sbs ← SBS (CCV)
       for each a[i][j] in A
                 if (sbs ∈ a[i][j])
                          w[i][j] = 0
       SBSᵢ ← sbs
return (SBS*)
```

Fig. 7. Pseudocode of the utility-aware redundancy minimization algorithm

6   Experimental Evaluation

In this section, we perform a couple of experiments to analyze the spatiotemporal coverage and the computational performance of our proposed algorithms on a real-world data set. In this case study, we focus on the SCOUTS project, which aims to generate hyperlocal heatmaps of the urban areas with high spatiotemporal granularity. The data set includes one-year trajectory data of twenty city buses in the city of Athens, Georgia.

6.1   Experimental Setup

To solve the coverage maximization problem of city buses, we create a grid covering the whole area while each cell corresponds to a 90-m by 90-m area on earth. In this study, the hotspot locations and their respective weights are determined by the heatmaps generated from satellite imagery. In other terms, the relative importance of each hotspot to be targeted by drive-by sensing vehicles is defined by analyzing the history of heatmaps generated by Landsat 8 satellite imagery.

We tested our proposed algorithms on 5 h of bus trajectory data collected every 5 s, which in total contains more than 61,000 data points. Furthermore, we located seven different hotspots in the city of Athens, with their weights varying from 2 to 8, provided by remote sensing experts. It should be mentioned that the hotspot locations in this experimental study, covered less than 0.075% of the whole urban area.

6.2   Results

In order to compare the coverage values along with the runtime performance of our three algorithms, we tested them on a varying number of sensors (i.e., 3, 4, 5, and 6 sensors representing 15% to 30% of the number of buses). Figure 8 shows the CCV earned by each algorithm and Fig. 9 depicts the runtime comparison of these algorithms. We can see that the utility-aware redundancy minimization algorithm consistently gains a CCV very close to the highest possible coverage value, while its runtime is orders of magnitude less than the exhaustive algorithm. In other terms, although the exhaustive algorithm guarantees to choose the bus combination with the highest possible CCV, its substantial computational cost, which grows significantly, makes it impractical for real-world applications. The considerable performance of our redundancy minimization algorithm is followed by the hotspot-based algorithm.

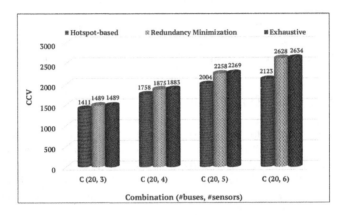

Fig. 8. CCV comparison for different number of sensors

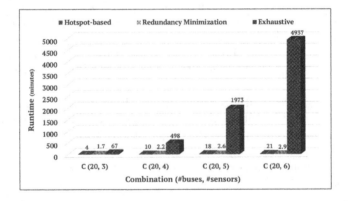

Fig. 9. Runtime comparison for different number of sensors

Finally, to elaborate on the practicality of our proposed objective function, we plot the trajectory data of a selected bus set (the experiment with 3 sensors)

on the map. Figure 10 illustrates how well our proposed objective function was able to select buses with the highest spatiotemporal coverage while having the lowest amount of overlaps.

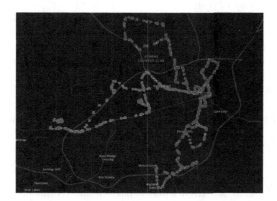

Fig. 10. Trajectory map of the bus combination of the exhaustive algorithm

7   Conclusion

In this paper, we first formulate the problem of choosing an optimal subset of non-dedicated mobile agents on which a limited number of sensors are to be mounted, for the sake of sensing coverage enhancement. Our objective function is implemented by three different algorithms, namely the exhaustive algorithm, the hotspot-based algorithm, and the utility-aware redundancy minimization algorithm. Then, we compare their performances and provide experimental results using real trajectory data set of public transportation buses in the city of Athens, GA. We showed how our utility-aware algorithms provide near-optimal solutions and outperform the exhaustive algorithm in terms of runtime. Our redundancy minimization algorithm is particularly practical for real-world drive-by sensing platforms where there are some regions with higher relative importance to be consistently monitored.

References

1. Tonekaboni, N.H., Ramaswamy, L., Mishra, D., Grundstein, A., Kulkarni, S., Yin, Y.: SCOUTS: a smart community centric urban heat monitoring framework. In: Proceedings of the 1st ACM SIGSPATIAL Workshop on Advances on Resilient and Intelligent Cities, pp. 27–30. ACM (2018)
2. Lee, U., Gerla, M.: A survey of urban vehicular sensing platforms. Comput. Netw. **54**(4), 527–544 (2010)
3. Anguelov, D., et al.: Google street view: capturing the world at street level. Computer **43**(6), 32–38 (2010)

4. Wang, M., Birken, R., Shamsabadi, S.S.: Framework and implementation of a continuous network-wide health monitoring system for roadways. In: Nondestructive Characterization for Composite Materials, Aerospace Engineering, Civil Infrastructure, and Homeland Security 2014, vol. 9063. International Society for Optics and Photonics, p. 90630H (2014)

5. Anjomshoaa, A., Duarte, F., Rennings, D., Matarazzo, T.J., deSouza, P., Ratti, C.: City scanner: building and scheduling a mobile sensing platform for smart city services. IEEE Internet Things J. **5**(6), 4567–4579 (2018)

6. Guo, B., Liu, Y., Wu, W., Yu, Z., Han, Q.: ActiveCrowd: a framework for optimized multitask allocation in mobile crowdsensing systems. IEEE Trans. Human Mach. Syst. **47**(3), 392–403 (2016)

7. Engelbrecht, A.: Particle swarm optimization. In: Proceedings of the Companion Publication of the 2014 Annual Conference on Genetic and Evolutionary Computation, pp. 381–406. ACM (2014)

8. Campioni, F., Choudhury, S., Salomaa, K., Akl, S.G.: Improved recruitment algorithms for vehicular crowdsensing networks. IEEE Trans. Veh. Technol. **68**(2), 1198–1207 (2018)

9. He, Z., Cao, J., Liu, X.: High quality participant recruitment in vehicle-based crowdsourcing using predictable mobility. In: 2015 IEEE Conference on Computer Communications (INFOCOM), pp. 2542–2550. IEEE (2015)

10. Yi, K., Du, R., Liu, L., Chen, Q., Gao, K.: Fast participant recruitment algorithm for large-scale vehicle-based mobile crowd sensing. Pervasive Mob. Comput. **38**, 188–199 (2017)

11. Wang, C., Li, C., Qin, C., Wang, W., Li, X.: Maximizing spatial-temporal coverage in mobile crowd-sensing based on public transports with predictable trajectory. Int. J. Distrib. Sens. Netw. **14**(8), 1550147718795351 (2018)

12. Dawson, C.R.: Heavy rail transit ridership report, first quarter 2009. American Public Transportation Association, pp. 08-16 (2009). apta.com. http://www.apta.com/resources/statistics/Documents/Ridership/2009_q2_ridership_APTA.pdf

13. Mediabridge Infosystems: The New York City bus system (2019). https://www.ny.com/transportation/buses/. Accessed 4 July 2019

14. Wmata Metro Service: Washington metropolitan area transit authority (2019). https://www.wmata.com/service/bus/. Accessed 4 July 2019

15. Roman, A.: Top 100 transit bus fleets: despite stimulus, funding issues remain for top 100, Metro, vol. 105, no. 8 (2009)

16. Mansoor, H.: Karachi needs 8,000 new buses to ease transport woes, CM told (2017). https://www.dawn.com/news/1323617. Accessed 4 July 2019

17. Murray, A.G.: World Trolleybus Encyclopaedia. Trolleybooks, Hampshire (2000)

18. Pirsiavash, H., Ramanan, D., Fowlkes, C.C.: Globally-optimal greedy algorithms for tracking a variable number of objects. In: CVPR. IEEE 2011, pp. 1201–1208 (2011)

Development of an Electronic System for the Analysis and Integration of Data on Water Care

Febe Hernandez Aleman$^{(\boxtimes)}$ and Martha S. Lopez-de la Fuente

Universidad de Monterrey, Nuevo Leon, Mexico
{febe.hernandez,martha.lopez}@udem.edu

Abstract. This work exposes the development of an electronic system for the detection of pollutant agents in water and integrates the design of the system with methods for collection of information and the detection of polluting particles in deposits. For this purpose, the stimulation is applied by dielectrophoresis, which manipulates the particles suspended in water, so that it is possible to perform a detection, extraction and accounting process to obtain data of the contaminants by doing experiment on liquid samples. The ultimate purpose is the anticipated detection of possible problems related to contaminants in the water.

The finished design and the functional prototype show experiments in the detection of live and inert particles in the range of 5 μm and 10 μm, and the methodology to carry out the proposed analysis. The elements of the system integrate the use of an LM3S6965 processor, which is responsible for processing the input parameters that pass through a signal conditioner, also it is responsible for the electrical stimulation necessary for the experiment and then using a microscope it's possible to generate images of the results. For the experiments shown there is the generation of signals of 5 Vpp with a frequency of 28 kHz, applied to a mixture of 10 μm polystyrene particles and 5 μm yeast cells, obtaining results about the behavior of these particles.

The main objective of the research carried out is to develop a solution to the problem that exists with the current methods that are not portable and to implement a system based on the System on Chip (SoC) architecture. The results of this work presents a functional prototype of the system, and shows the detection of contaminants and a data transmitter for the collection of information. Also, the experiments made show the handling of inert and living particles with the use of polystyrene and yeast beads. One of the main characteristics of the system is that it could be adjusted to read parameters in other liquid containers to detect different particles.

Keywords: Particle manipulation · Dielectrophoresis · Pollution on water deposits · System on chip

1 Introduction

For the present research, emphasis will be placed on presenting a system capable of identifying different types of particles in static aquatic environments, that is, in water

© Springer Nature Switzerland AG 2020
W. Song et al. (Eds.): ICIOT 2020, LNCS 12405, pp. 125–132, 2020.
https://doi.org/10.1007/978-3-030-59615-6_9

containers, which could contain particles or contaminating agents. Among the different types of microorganisms that can exist in aquatic containers, are include bacteria, fungi, protozoa, arthropods, among many others. Although some of these microorganisms could be useful to maintain balance in aquatic environments, some of them can be harmful to health and represent problems in the quality of the water that contains them [1].

In this way, there are different factors that can alter the healthy parameters of the water contained in different deposits. On the other hand, the terms and criteria that exist for the care of water quality are applied primarily to human consumption, followed by deposits that represent habitats for wildlife and applications such as recreational and industrial uses [2]. Therefore, in order to comply with the required standards, it is necessary to perform tests, experiments and measurements.

There are different types of tests that can be performed on water containers. According to the World Health Organization, water with polluting agents can be transmitted as cholera, typhoid fever, poliomyelitis, among others; and at the same time, it is estimated that the pollution on drinking water causes more than 502,000 deaths per year [3]. In this way, conducting research that promotes innovation in systems for the analysis of water tanks and that at the same time are simple and accessible to end users, represents a latent need.

Studies on particle handling impact a wide range of research areas and have different applications. Also, there are different methods that allow these analyzes to be carried out, to explore new phenomena in the mentioned areas and to open the way to new experiments [4]. On the other hand, one of the most used methods to carry out analyzes of this type is the stimulation of particles in an electric way, however, one of the limitations that the use of this type of methods has is that the instruments used are not portable, since they depend on the amplitude of the signals, as well as on the frequency used to generate the desired effect in the analyzed particles [5].

For the purpose of this study, a particle analysis in water tanks will be carried out, implementing dielectrophoresis methods in order to detect contaminating agents in the mentioned tanks. Dielectrophoresis is defined as a process of particle transport that is possible through the application of non-uniform electric field [6]. Said method consists of applying electric fields in particles suspended in some fluid, in this case, the polluting agents that are suspended in the water tanks. When this force is applied to the particles, the respective elements experiences an imbalance internally, causing the movement of the particles.

2   Materials and Methods

2.1   Particle Manipulation

The studies in the area of particle handling have been of great help in the development of new systems that allow improving the quality of people's lives. These significant advances have made evident the need to continue innovating in the development of new techniques or in the improvement of existing ones.

The electromechanical phenomenon of inducing force to suspended particles includes concepts such as electro-osmosis (EO), electrophoresis (EP) and dielec-trophoresis (DEP), among others [7]. This type of methods has been useful in the development of techniques that allow concentrating and separating particles for their study. To continue with the objective of this research, the technique chosen as the object of study is Dielectrophoresis (DEP). In this way, the magnitude of the force applied must be a function of the radius of the particle (r) and the gradient of the electric field squared ($\nabla E2$). The polarization of the suspended particle is produced by the electri-cal properties, permittivity and conductivity, as well as by the frequency of the applied signal, and is defined by the Clausius-Mossotti factor. This factor is recurrently used in studies involving Dielectrophoresis, because it allows recognizing the polarization of the particle embedded in a specific container. The force of Dielectrophoresis will be positive (attractive) when the particle is polarized to a greater extent than the suspension liquid, presenting an attraction to the regions where the intensity of the electric field is high and the permissiveness is higher than the mean respectively; otherwise, the force will have a negative (repulsive) magnitude. The following equations show how to obtain the force of Dielectrophoresis and how to calculate the Mossotti Factor.

$$\overrightarrow{F_{DEP}} = 2\pi r^3 \varepsilon_0 \varepsilon_m Re(f_{CM})\Delta E^2 \tag{1}$$

$$f_{CM} = \frac{\varepsilon_p - \varepsilon_m}{\varepsilon_p + 2\varepsilon_m} \tag{2}$$

There are different causes why an experiment with Dielectrophoresis can be affected. Variables such as the size and shape of the molecules of interest, the temperature, the applied voltage and the duration of the applied electrical force, just to mention a few. In this way, in order to reduce such variations in experiments of the same type, there are parameters that function as limits depending on the type of material used in the experiment.

2.2 System Design

The design of the presented system consists of the application of a System-on-Chip (SoC), with the purpose of modeling and implementing electrical signals for the manip-ulation of the studied particles. In the same way, the design includes a transmitter module for collecting data on the experiments carried out. In this way, the implementation shows the configurable system, capable of generating individual, double or superposed signals of up to 30 Vpp as output, said signals can be sinusoidal, saw, or triangular. Also, the system can use frequencies in a range from 0.01 kHz to 40 kHz. One of the main char-acteristics of the presented system is that it can be reprogrammed and configured for the generation of different types of signals. In the following figure, each element of the system is specified and the operation of each one is mentioned.

As shown in Fig. 1, the serial connection of each of the elements of the system is done through a USB port that connects the user interface with the processor for signal stimulation.

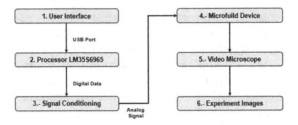

Fig. 1. Flow chart with the elements of the system

The interface of the system has been developed in order to provide the user with control over the parameters entered into the system. Said operating parameters are the type of signal that will model the system, the frequency of said signal, the samples per cycle and the period time for the simulation when it is not a continuous cycle experiment. The interface is displayed on the computer screen where the measuring system is connected via USB. The next step is for the user to enter if he wants to perform an experiment with a single signal, two signals on two different channels, or two superimposed signals on the same channel.

2.3 Signal Conditioning

The signal conditioner is necessary to transform the digital data coming from the processor into an analog signal. To make this possible, the DAC902 device is used, which is a digital-analog converter, as well as two AD811, which are amplifiers and allow to manipulate the characteristics of the signals as they are necessary for the application. This device delivers as an output parameter an analog signal that goes from 0 to 30 Vpp. This signal complies with the parameters of a classical signal used in experiments on particle handling. In the following group of figures, the different outputs that the device used can have are shown.

As shown in Fig. 2a to 2f the device presented in this work can generate signals of saw, sinusoidal, triangular type, and even some combinations between sinusoidal and triangular, or triangular with saw with different voltage levels, which can range from 0 to 30 Vpp.

2.4 Particle Stimulation and Detection

For the presented system, a CarbonDEP device was used, which has three-dimensional electrodes. These electrodes are necessary to apply the voltage produced by the non-uniform electric field, which generates the necessary dielectrophoresis effect. For the experiments carried out in the presented research, yeast particles were used, in order to carry out the necessary tests and to prove the correct functioning of the signal generator. In addition, it was necessary to implement an arrangement with three carbon electrodes, which are used to apply the electric potential necessary to produce a non-uniform electric field, with which it is possible to generate the necessary dielectrophoresis force for the experiment.

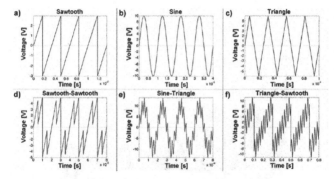

Fig. 2. **a – f** Output signals

In Fig. 3, a graphic representation of the arrangement of the carbon electrodes can be seen. The negative electrodes are represented in dark blue, while the positive electrodes are represented in light blue.

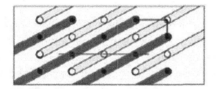

Fig. 3. Graphic representation of the carbon electrodes array (Color figure online)

On the other hand, the three-dimensional structure of carbon is 40 μm in height, with a radius of 12.5 μm, and a center-to-center separation of 45 μm on the X axis and 100 μm on the Y axis. Likewise, deionized water was used with K2HPO as a container and suspension medium for the particles. The conductivity obtained with the carbon arrangement and the liquid used is 21 μS/cm. In this case, the detection of the particles can be done in two different ways; the first, using a resistive sensor which measures the changes between two electrodes, caused by the movement of the particles; and the second, applied in the experiments carried out, using a camera and a microscope.

3   Results

3.1   Prototype Implementation

In Fig. 4, the physical appearance of the system is shown, in the same way, each of the elements mentioned above is indicated in the flow chart of Fig. 1. For the selected experiment, an analog signal of 5 Vpp with a frequency of 28 kHz was produced, and a mixture of 10 μm polystyrene particles and 5 μm yeast cells was selected. In the experiment performed, the different particles showed positive and negative Dielectrophoresis behaviors. On the other hand, estimates of the force Dielectrophoresis made using the application

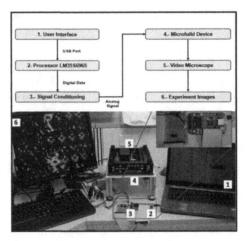

Fig. 4. Flow chart and physical representation of the elements of the system

of a finite element method. For this, a 1.2 mm wide microchannel was considered, which contains an array of 4 × 5 electrodes.

When the system is working, in first place, the processor (Number 2, Fig. 4) takes the data entered and processes it in the signal conditioning module (Number 3, Fig. 4), then transport the data in the form of the desired analog signal. Then takes the image with the video microscope (Number 5, Fig. 4) and displays it on the computer (Number 6, Fig. 4). The last part of the experiment is analyze the images.

Thus, for the estimates made, it was considered for a plane located 30 μm above the channel floor. At this height, the effect of the flat electrodes located in the lower part of the channel could be ignored. The boundary conditions were established in electrical isolation in the walls of the channel and uniform AC electric potential in the three-dimensional electrode poles. The mesh for this geometry consisted of 14,208 elements. Also, it can be seen that, for the frequency range considered in this experiment, the polystyrene beads will experience negative DEP, while the yeast cells present a change in 8 responses, changing from weak DEP at low frequencies to strong DEP at high frequencies. Experimental results for the mixture of yeast cells and 10 μm polystyrene particles are presented in Fig. 5b and 5c. In Fig. 5c it is observed that, when no potential is applied, the particles are distributed randomly after injecting the sample into the channel.

The experimental results confirm that the yeast cells show a positive Dielectrophoric force response that attracts carbon structures and flat carbon electrodes in the lower part of the channel. Meanwhile, polystyrene beads show a negative response to the force of Dielectrophoresis generated, since it repels carbon electrodes and flat electrodes in regions where the gradient is lower between the structures.

3.2 Discussion

The experiments were performed with polystyrene and yeast particles, however, the developed system can be installed in any water tank and applied to identify a wide variety of different particles. In a study where the design of a device with an arrangement of

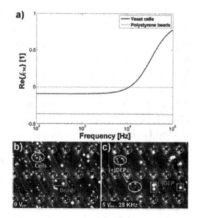

Fig. 5. a–c Clausius Mossotti factor and results of the mixture of yeast and polystyrene particles

insulators inside a channel was made, where they managed to trap latex particles of 200 mm in diameter; a technique similar to that used in the present investigation.

Regarding the use of the Clausius Mossotti Factor, a study carried out a work in which it's introduced a model for the modeling of dielectrophoresis using said factor [8]. This analysis illustrates how to add frequencies and parameters to obtain particle separation with a similar Clausius Mossotti factor. The results obtained in said study show the use of different frequencies for the manipulation of particles; a process similar to that presented in the current investigation.

Therefore, methodologies that use the use of dielectrophoresis will take even greater importance when dealing with issues such as water quality or public health, since they involve the early detection of diseases or have the opportunity to avoid some of them through the detection of particles harmful in time.

3.3 Conclusions

A system with these characteristics opens the potential for more complex systems for the manipulation of particles in microfluidic devices, different stages could be integrated in a simple device, since each stage can have an independent and different electrical stimulation. Complex samples can be handled in steps, from the initial classification, followed by separation and concentration. Also, the system could be used not only in water deposits but in other type of liquids, opening the opportunities to improve quality in different situations. In this way, the versatility of the device allows it to be used in a wide range of applications.

And since the system is portable, a sequence of experiments can be performed without the need of user intervention. Finally, the system is also capable of working with non-sinusoidal waveforms, such as those of triangular or saw teeth, it provides an additional degree of flexibility for the development of new electrokinetic manipulation techniques based on the combination of these signals. The system described here is a step towards the development of more advanced on-chip integrated systems.

References

1. Boyd, C.E.: Water Quality. Springer, Cham (2015). https://doi.org/10.1007/978-3-319-17446-4
2. Arroyo, M.: Importancia de la calidad del agua y su manejo (2011). https://ucienegam.mx/wp-content/uploads/2017/08-Doc/Servicios%20Escolares/Alumnos/Optativas-Febrero/Importancia_de_la_calidad_del_agua_y_su_manejo.pdf
3. OMS-Agua (n.f.). Datos y Cifras. Consultado el 11 de marzo de 2019. https://www.who.int/es/news-room/fact-sheets/detail/drinking-water
4. Salmanzadeh, A., et al.: Lab Chip **12**, 182 (2012)
5. Jesús-Pérez, N.M., Lapizco-Encinas, B.H.: Electrophoresis **32**, 2331 (2011)
6. Gallo Villanueva, R.C., Lapizco Encinas, B.H.: Dielectroforesis para sistemas micro-escala. Espacio del Divulgador. Volumen **20**, 227–230 (2013)
7. Lopez-de la Fuente, M.S., Moncada-Hernandez, H.: An Electric stimulation system for Electrokinetic Particle Manipulation in Microfluidic Devices, American Physics Association, Review of Scientific Instruments vol. 84 (2013). https://doi.org/10.1063/1.4793559
8. Urdaneta, M., Smela, E.: Electrophoresis **28**, 3145 (2007)

BWCNN: Blink to Word, a Real-Time Convolutional Neural Network Approach

Albara Ah Ramli[1(✉)], Rex Liu[1], Rahul Krishnamoorthy[2], I. B. Vishal[2], Xiaoxiao Wang[1], Ilias Tagkopoulos[1], and Xin Liu[1]

[1] Department of Computer Science, University of California, Davis, CA, USA
{arramli,rexliu,xxwa,iliast,xinliu}@ucdavis.edu
[2] Electrical and Computer Engineering, University of California, Davis, CA, USA
{rkrishnamoorthy,vib}@ucdavis.edu

Abstract. Amyotrophic lateral sclerosis (ALS) is a progressive neurodegenerative disease of the brain and the spinal cord, which leads to paralysis of motor functions. Patients retain their ability to blink, which can be used for communication. Here, We present an Artificial Intelligence (AI) system that uses eye-blinks to communicate with the outside world, running on real-time Internet-of-Things (IoT) devices. The system uses a Convolutional Neural Network (CNN) to find the blinking pattern, which is defined as a series of Open and Closed states. Each pattern is mapped to a collection of words that manifest the patient's intent. To investigate the best trade-off between accuracy and latency, we investigated several Convolutional Network architectures, such as ResNet, SqueezeNet, DenseNet, and InceptionV3, and evaluated their performance. We found that the InceptionV3 architecture, after hyper-parameter fine-tuning on the specific task led to the best performance with an accuracy of 99.20% and 94 ms latency. This work demonstrates how the latest advances in deep learning architectures can be adapted for clinical systems that ameliorate the patient's quality of life regardless of the point-of-care.

Keywords: CNN · IoT · Neural Network · Transfer learning · Resnet · Inception · InceptionV3 · DenseNet · Squeezenet · ALS

1 Introduction

A plethora of clinical conditions, such as brain trauma and amyotrophic lateral sclerosis (ALS), cause damage to the central neural system (CNS) or brain, in such as way that the ability of speech and motor functions cannot be sustained. In those cases, the ability to communicate is limited, to non-verbal forms of communication, such as eye blinking.

Certain approaches have been used in the past to solve this problem. Researchers in [1] use Infrared (IR) sensors to estimate the state of the eyes ('Open' or 'Closed') in order to detect blinking, which is then converted to Morse code. Challenges in this approach include the IR sensor being irradiated

© Springer Nature Switzerland AG 2020
W. Song et al. (Eds.): ICIOT 2020, LNCS 12405, pp. 133–140, 2020.
https://doi.org/10.1007/978-3-030-59615-6_10

by other sources resulting in false eye-blinks, and the risk of cataract forma-
tion in the case of prolonged use [1]. In another study, a traditional computer
vision-based system that detects users' eye-blinks, measures their duration, and
interprets the blinks in real-time was proposed [2]. Although the system has
achieved an accuracy of 95.6% in ideal conditions, traditional image processing techniques are prone to failure under limited lighting conditions, arbitrary
changes in image texture, changes in user s pose, among others, which limits
their real-time accuracy, and make these systems not robust in real applications.

A wearable device to detect eye-blinks for alleviating dry eyes was proposed
in [7]. This method captured 85.2% of all the blinks that occurred during testing.
But, the IR sensors tend to show false readings when the orientations are altered
and hence, are unreliable in realistic scenarios. Any facial movements such as
laughing, or yawning can induce errors.

To avoid these shortcomings, we here present a deep-learning vision system
that detects eye-blinks and maps them to words in real-time. Our system uses the
InceptionV3 [12] architecture and achieves an accuracy of 99.20% with a latency
of 94.1 ms on IoT devices. Our method is safe to use, has better performance
than previous methods, it is robust to changes in lighting conditions and facial
orientation, and its architecture is modular so its output can be mapped to other
tasks, such as controlling software applications or devices. The main contribution
of this paper resides in designing, implementing, and evaluating the first deep-
learning solution for eye-blink communication with performance, latency, and
safety specifications that can be used in a real-world environment.

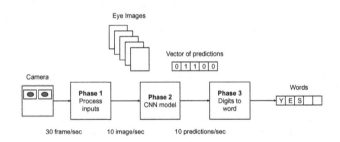

Fig. 1. The 3 phases of the BWCNN system.

1.1 Previous Work

An efficient system for eye-blink detection is presented in [3]. This method uses
Haar-cascade classifiers for face detection and eye positions. The performance
of Haar-cascade classifiers is not invariant to the change in lighting conditions.
Hence there is a decay in performance.

A low-cost implementation of an eye-blink-based communication aid for ALS
patients is presented in [4]. Template matching is used to track the eye and detect
eye-blinks using hierarchical optical flow. The implementation has an accuracy

of 94.75% during the typing test. However, the algorithm takes approximately 2 s to generate a single scan of the eye. This is excessive for a single character.

This paper [6] presents a real-time detection and classification between eye-blink (with both eyes), left wink, or right wink with 96, 92, and 88% accuracy. The latency for the detection of a single blink was 250 ms.

A wearable device to detect eye-blinks for alleviating dry eyes was proposed in [7]. This method captured 85.2% of all the blinks that occurred during testing. But, the IR sensors tend to show false readings when the orientations are altered and hence, are unreliable in realistic scenarios. Any facial movements such as laughing, or yawning can induce errors.

2   Design and Implementation

Our system detects the state of eyes, 'Open' and 'Closed', even under poor lighting conditions. We have a pre-defined set of patient inputs corresponding to the blink pattern, which we map to actions in real-time. These inputs could correspond to movements (up, down, left, right), clicks, etc., which would enable patients to use different applications (browser, email, etc.) or devices (mouse, keyboard, etc.). As a proof-of-concept, we mapped these inputs to specific words. Since we use predefined words instead of Morse code or other encoding patterns, it is simple for patients to spell out a sentence. We want to create a system that works almost flawlessly in real-time and is safe to use. This is represented by equation (1). $\mathcal{P}$ refers to the performance of the system, which in our case is the accuracy. $\mathcal{S}$ refers to the System parameters. $\mathcal{W}$ refers to the weights of the neural network. $\mathcal{A}$ refers to the architecture. $\mathcal{H}$ is the set of all architectures that can be used for this purpose. $\mathcal{T}$ refers to time constraint which is the prediction time (response time, or latency) for the model on the validation set. It is important to reduce the response time of the system since it will be running over a real-time IoT device such as Raspberry Pi.

$$\max_{\mathcal{A}\in\mathcal{H}} \mathcal{P}(\mathcal{S}, \mathcal{W}, \mathcal{A})$$

$$s.t \; prediction \; time \leq \mathcal{T}$$

(1)

The goal of this approach is to detect if the patients blink their eyes and to map the sequence of blinks to a particular entry in a dictionary of words. In order to achieve this, one has to detect the state of eyes ('Open' or 'Closed'). If an 'Open' state is followed by a 'Closed state, the system detects an eye-blink, as shown in Fig. 2. The system is divided into three phases as shown in Fig. 1.

2.1   Phase 1: Capturing and Saving a Stream of Frames

In this section, we present the method of obtaining the data and preprocessing it to be given as input to the ConvNet. The system uses a camera device attached to IoT for capturing the frames. Regular webcams are capable of capturing about

30 frames per second (fps). The fps will directly affect the user experience. A higher fps will increase the latency.

Since the system runs in real-time, it is more effective to reduce the latency. At the same time, using lower fps can lead to missing a blink. Our experiment shows that 10 frames per second are a reasonable frame rate for real-time application. As there is a frame being captured every 100 ms, our model must predict each frame in less than 100 ms to avoid delay. We further impose a constraint on the user that the 'Closed' state should be maintained for at least 200 ms. The system saves each frame as a gray-scale image of dimensions 80×70 pixels. Each image is 2 KB in size.

2.2 Phase 2: Predict the Content of the Image

In this section, we present the experiments to choose the best fitting neural network architecture to predict the state of the eye. We compare four state-of-the-art architectures (SqueezeNet [5,10], ResNet [8], InceptionV3 [11] and DenseNet [9] architecture). Both the architecture and the hyperparameters play a large role in model performance. We start by training the networks from scratch for different batch sizes and find the batch size that gives the best results. We further explore the chances of improving performance by using transfer learning.

Training Dataset: We used the eye dataset from Media Research Lab (MRL). The dataset contains 84,898 images of eyes taken from 37 individuals consisting of 33 men and 4 women. Each image in the dataset was collected from one of the following sensors: Intel RealSense RS 300 sensor with a resolution of 640×480, IDS Imaging sensor with a resolution of 1280×1024, and Aptina sensor with a resolution of 752×480. The original dataset contains 6 classes: 'gender', 'glasses', 'eye state', 'reflections', 'lighting conditions', and 'sensor resolution'. We split the dataset into an 80% training set and a 20% test set.

Training Experiments: Two important considerations when training the model were accuracy and latency. Latency for detection is the time taken to make an accurate classification. To find the model which can provide the best accuracy with the least latency, we implemented SqueezeNet, ResNet, InceptionV3, and DenseNet.

Train ResNet Architecture from Scratch: We trained the ResNet architecture for 100 epochs using 6 batch sizes. The performance metric used is the overall accuracy. Comparing the performance, we selected the three best batch sizes from 1, 2, 4, 8, 16, and 32. Table 1 shows that batch sizes 8, 16, and 32 provide the best accuracy. Since our experiment stops at 100 epochs, training the network for more epochs might improve the performance. To test this, we ran ResNet, for all batch sizes for 500 epochs. The results show that there is no further improvement in accuracy.

Table 1. ResNet with and without transfer learning

Batch S.	Epoch	Training a model (from scratch)		Transfer learning (from our model)		Transfer learning (from official)	
		Acc. (%)	Ep. imp.	Acc. (%)	Ep. imp.	Acc. (%)	Ep. imp.
8	100	99.21	32	99.22	31	99.22	29
16	100	99.26	55	99.23	51	99.17	16
32	100	99.22	48	99.22	49	99.17	25

Table 2. DenseNet, SqueezeNet, InceptionV3

Batch S.	Eep.	DenseNet		SqueezeNet		InceptionV3	
		Acc. (%)	Ep. imp.	Acc. (%)	Ep. imp.	Acc. (%)	Ep. imp.
8	100	99.24	55	49.40	1	99.14	35
16	100	99.18	70	49.40	1	99.20	22
32	100	99.21	52	49.40	1	99.17	38

Transfer Learning: We took the weights from the most accurate architecture, ResNet with batch size 16, and transferred these weights to all other ResNet architecture with the 3 best performing batch sizes. We also did transfer learning by using weights from the official pre-trained ResNet on our ResNet models with the 3 best performing batch sizes. As before, after 100 epochs, there was no significant improvement in the accuracy of the network.

Train InceptionV3, SqueezeNet and DeneseNet Architectures from Scratch: We assume that the 3 best-performing batch sizes from ResNet would be the best performing in the other architectures as well. To investigate this assumption we run the same experiment again but with DenseNet, Inception, and SqueezeNet as shown in Table 2.

2.3   Phase 3: Mapping

Our system stored the output of the neural network as a vector of 0s and 1s, where Zero represents the 'Open' state for the eye and a One represents the 'Closed' state as shown in Fig. 2. We normalize the vector by truncating repeated instances of a state with a single instance. For example, vector 00000110000 becomes 010. The system provides a special word to start and end a session of blinking, which is keeping the eyes closed for 4 s. The system recognizes the end of a word when the patient's eyes remain open for 1 s. Based on the output vector, the number of blinks is calculated and is mapped with the words in the dictionary (Table 3). The dictionary consists of basic words that we use as a proof-of-concept and it can be modified.

Table 3. Dictionary

# blinks	1	2	3	4	5	6	7
Pattern	1	101	10101	1010101	101010101	10101010101	1010101010101
Words	Yes	No	Hi	I am	Good	Thanks	How are you?
Mouse	Right	Left	Click R.	Click L.	Up	Down	Hold click
Keyboard	Tab	Enter	Back	Esc	Scroll up	Scroll down	Space

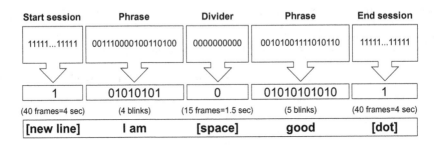

Fig. 2. BWCNN system predicting a series of frames in the real-time

3　Results

This section provides the performance results of the different architectures. Table 1 lists the results obtained from training ResNet for 100 epochs on batch sizes 1, 2, 4, 8, 16, and 32. Training ResNet from scratch for 100 epochs on a batch size of 16 yields the best accuracy, 99.26%. There is no improvement in accuracy after 55 epochs (Fig. 3).

Table 4. Final results: 3.1 GHz Quad-Core Intel Core i7, 16 GB 2,133 MHz

Model architecture	Batch size	Ep. imp.	Total params	Trainable params	Nontrainable params	Model size	Model accuracy	Avg latency
ResNet	16	55	23,591,810	23,538,690	53,120	283 MB	99.26%	117.28 ms
DenseNet	8	55	7,039,554	6,955,906	83,648	85 MB	99.24%	146.09 ms
SqueezeNet	16	1	723,522	723,522	0	8 MB	49.40%	13.64 ms
InceptionV3	16	16	21,806,882	1,772,450	34,432	262 MB	99.20%	94.1 ms

We trained a new ResNet model using transfer learning. We took the weights from the best performing network that we had trained form scratch and used them to train a ResNet network with batch sizes of 8, 16, and 32. We also did transfer learning with weights obtained from the official pre-trained ResNet architecture and used it to train ResNet architecture with batch sizes of 8, 16, and 32. The results are shown in Table 1.

Table 2 presents the results for DenseNet, SqueezeNet, and InceptionV3 architectures for the batch sizes of 8, 16, and 32. Figure 4 shows the training and loss curve for InceptionV3. DenseNet has the best accuracy of 99.24%,

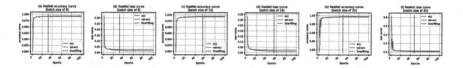

Fig. 3. ResNet: training and validation (accuracy and loss) curves.

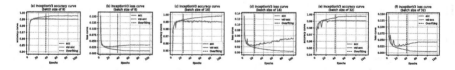

Fig. 4. InceptionV3: training and validation (accuracy and loss) curves.

which is achieved for a batch size of 8, as training for 55 epochs, after which it starts over-fitting.

SqueezeNet seems to be the worst among all the architectures with an accuracy of 49.40%. There is no change in accuracy after the first epoch, which is in Table 2. The final comparison of the best results across the different architectures in Table 4. Apart from the accuracy of the networks, Table 4 also contains the latency (in milliseconds). We can see that ResNet has the best accuracy of 99.26%, but has a high latency of 117.28 ms. InceptionV3 has an accuracy of 99.20% which is close to ResNet, but with a lower latency of 94.1 ms. We can see that InceptionV3, DenseNet, and ResNet have similar accuracies but the InceptionV3 model has the lowest latency. Thus, we chose InceptionV3 architecture.

4   Conclusion

In this paper, we designed an AI system for non-verbal communication that converts eye-blinks to words using a deep learning CNN architecture. The system predicts the state of the eyes of the patient and finds the blinking pattern. We compared several CNN architectures and hyperparameter selections in model training. For the evaluation, we tested our system using 16,979 facial images and found that our proposed prediction model was efficient and effective. Results demonstrate that overall prediction accuracy is 99.20% and the average prediction time is 94 ms. We trained different architectures with different hyperparameters to identify parameter combinations that lead to high accuracy and low latency. For the sake of conducting a clear comparative analysis, we compare the results of each architecture with batch sizes of 8, 16, and 32. SqueezeNet received the lowest accuracy with the fewest parameters. The DenseNet, ResNet, and InceptionV3 acquired accuracies in the range of 99.20% and above. As InceptionV3 had the lowest latency, we chose this architecture. We introduced transfer learning, which improved the convergence when compared to random initialization, to otherwise similar accuracy and latency in the response. Future work includes the training on a more generalized training dataset, the application of hybrid technologies that fuse computer vision techniques such as the one presented together with other natural language processing methods, including recurrent neural networks, to introduce memory in the actions and system. The dataset, source code, demo, and results of this system are available at https://albara.ramli.net/research/bwcnn/.

References

1. Mukherjee, K., Chatterjee, D.: Augmentative and Alternative Communication device based on eye-blink detection and conversion to Morse-code to aid paralyzed individuals. In: 2015 International Conference on Communication, Information and Computing Technology (ICCICT), Mumbai, India, pp. 1–5 (2015)
2. Grauman, K., Betke, M., Gips, J., Bradski, G.R.: Communication via eye-blinks - detection and duration analysis in real-time. In: Proceedings of the 2001 IEEE Computer Society Conference on Computer Vision and Pattern Recognition, CVPR 2001, vol. 1, pp. I-I (2001)
3. Mohammed, A., Anwer, S.A.: Efficient Eye Blink Detection Method for disabled-helping domain. Int. J. Adv. Comput. Sci. Appl. **5**, 4–5 (2014)
4. Su, M.-C., Yeh, C., Lin, S., Wang, P., Hou, S.: An implementation of an eye-blink-based communication aid for people with severe disabilities. Presented at the ICALIP 2008–2008 International Conference on Audio, Language and Image Processing, Proceedings, pp. 351–356 (2008)
5. Iandola, F., Han, S., Moskewicz, M., Ashraf, K., Dally, W., Keutzer, K.: SqueezeNet: AlexNet-level accuracy with 50x fewer parameters and 0.5 MB model size, 1 (2016). arXiv preprint arXiv:1602.07360
6. Singh, Hari, Singh, Jaswinder: Real-time eye blink and wink detection for object selection in HCI systems. J. Multimodal User Interfaces **12**(1), 55–65 (2018). https://doi.org/10.1007/s12193-018-0261-7
7. Dementyev, A., Holz, C.: DualBlink: a wearable device to continuously detect, track, and actuate blinking for alleviating dry eyes and computer vision syndrome. In: Proceedings of the ACM on Interactive, Mobile, Wearable and Ubiquitous Technologies, vol. 1, no. 1, pp. 1–19, March 2017
8. He, K., Zhang, X., Ren, S., Sun, J.: Deep residual learning for image recognition. In: CVPR (2016). 2, 5, 6, 7
9. Huang, G., Liu, Z., Weinberger, K.Q., Maaten, L.: Densely connected convolutional networks. In: CVPR (2017). 2, 4, 6
10. Hu, J., Shen, L., Sun, G.: Squeeze-and-excitation networks. arXiv preprint arXiv:1709.01507 (2017). 1, 2, 5, 7
11. Szegedy, C., et al..: Going deeper with convolutions. In: CVPR (2015). 1, 2, 4
12. Szegedy, C., Vanhoucke, V., Ioffe, S., Shlens, J., Wojn, Z.: Rethinking the inception architecture for computer vision. In: CVPR (2016). 2, 4, 6

Risk Assessment of Vehicle Sensor Data as a Vending Object or Service

Frank Bodendorf[(⊠)] and Jörg Franke

Institute for Factory Automation and Production Systems, Friedrich-Alexander-University of
Erlangen-Nuremberg, Egerlandstraße 7-9, 91058 Erlangen, Germany
{frank.bodendorf,joerg.franke}@faps.fau.de

Abstract. Connected cars generate a huge amount of vehicle data during operation. In the future, the amount of sensor-generated data will continue to increase.
The connectivity of the cars, more powerful processors, and improved telematics
and navigation systems will allow the amount of data to grow further. Vehicle
data provides a basis for a large number of business models. In addition to selling
vehicles, automobile manufacturers can generate additional revenue by selling
vehicle-generated data as goods or services. First, a typology of vehicle data is
described in order to derive value potentials of data products. Motivated by the
value perspective, risks in data transfer to third parties are often neglected. In order
to assess these risks, a new risk management model for intangible products like
data is presented. The main phases of the risk assessment procedure are walked
through, outlining possible criteria and metrics in each phase. Finally, the model is
demonstrated by evaluating risks of data transfer to third parties in the automotive
industry, using the example of vehicle-generated road segment data.

Keywords: Connected car · Internet of Things · Cyber physical systems · Sensor
data · Data security · Business model · Risk management · Automotive industry

1 Introduction

Each new Internet wave has produced new business model patterns so far. Between the
years 1995 and 2000, business models such as e-commerce and open source emerged
in the wake of Web 1.0, i.e. the Internet as a business medium. With Web 2.0, which is
primarily characterized by the term "social media", e.g., crowdfunding and crowdsourcing emerged. It is assumed that in the course of Web 3.0, which is often associated with
the Internet of Things, the business model patterns "Digitally Charged Products" and
"Sensors as a Service" in particular will prevail [9]. The term "Digitally Charged Products" characterizes physical products enhanced by new service promises that include
data-based service bundles [9]. "Sensors as a Service" refers to the collection, processing, and resale of sensor data to third parties [23]. These digital business models of the
Internet of Things (IoT) become relevant not only for digital industries, but increasingly
also for physical industries like automotive manufacturing [9].

© Springer Nature Switzerland AG 2020
W. Song et al. (Eds.): ICIOT 2020, LNCS 12405, pp. 141–151, 2020.
https://doi.org/10.1007/978-3-030-59615-6_11

2   Research Motivation and Research Objectives

In the future, the amount of vehicle-generated data will continue to grow. More and more sensors are being installed, the connectivity of cars is increasing and more powerful processors and improved telematics and navigation systems are being used [29, 31]. The data generated by a vehicle is transmitted to the backend in several ways. On the one hand, original equipment manufacturers (OEMs) receive data via the onboard diagnostics interface when reading it out in a workshop or at a dealer. On the other hand, a lot of data is also transmitted in real time via the mobile phone network. This data is then prepared, aggregated, refined, analyzed, and combined with additionally available data (e.g., real-time traffic information). Based on this data, the car interacts with the physical and digital world by means of actuators [13, 25].

This vehicle-generated data is a very valuable asset and enables a variety of services that improve the driving experience. But in particular, they are also of great benefit to third parties (besides the driver and the vehicle manufacturer/OEM) [3]. In this context, keywords such as "Data as a Service" or "Sensors as a Service" have been coined.

Just like other business models of the Internet of Things data plays a crucial role in new service models for networked vehicles [17, 21, 28, 39]. The "Data Orchestrator" is seen as the most significant business model pattern in relation to connected cars [25]. The business model pattern "Leverage Customer Data" is also closely linked to the IoT business model pattern "Sensors as a Service" and deals with the collection, processing, and sale of customer and vehicle data. Within the framework of "Data Orchestrator", OEMs act as leaders of a platform that enables various service providers, e.g., from the software or electronics industry, to interact openly and take advantage of network effects that occur primarily in connection with data processing [25].

Data and data-driven services are of great importance for the Internet of Things in general and for networked cars and the automotive industry in particular [34]. However, these data-based services are associated with risks such as financial risks, macroeconomic risks, socio-cultural risks or political risks [6]. More specifically, these include risks such as the loss of data, the manipulation of data records or the failure of cloud services. Only a small number of publications on risk management of data-driven business models and services can be found in the scientific literature [7, 22, 27, 39]. In practice, however, the risk issue is highly relevant [14, 37].

On the one hand, data owners see many opportunities to generate additional revenue by selling the data as a marketable product (in raw or processed form) and by offering paid data-based services. On the other hand, the data owners or holders recognize that the selling of data as a product or service can be associated with high risks, e.g., of misuse or unauthorized proliferation. This paper presents an approach to assess risks of selling and transferring vehicle-generated data to third parties like service providers that use the data for their own business models. The following questions are addressed:

1. *Which aspects could a risk assessment model for vehicle-generated data take into account?*
2. *How does the application of this model look like in a practical example?*

Section 3 characterizes vehicle data generated by sensors. This is done from a technical as well as from a value oriented perspective. Section 4 presents a model created for risk assessment of data-driven services in the automotive industry. The focus is on risks of data transfer to third parties as buyers. The basic application of the model is illustrated by a use case referring to sensor generated road segment data. Finally, a critical evaluation of the approach is given in Sect. 5.

3 Sensor Systems and Valuable Data

The automotive industry, among others, is particularly affected by the digital transformation [30]. Cyber-physical systems play an increasingly important role in this industry [2, 20]. More and more automobiles are equipped with a huge number of sensors and connected to other vehicles and objects. They generate, process, combine, use, and share data with other devices and vehicles [26]. They are thus an important part of the Internet of Things [34]. In this way, the vehicle itself can be regarded as a cyber-physical system that consists of various networked subsystems [25]. A large part of the data is generated by the sensors and cameras of the driver assistance systems, which continuously scan the vehicle's environment [8, 10, 11, 33, 38]. Figure 1 shows different sensor fields of an assistance system (AS).

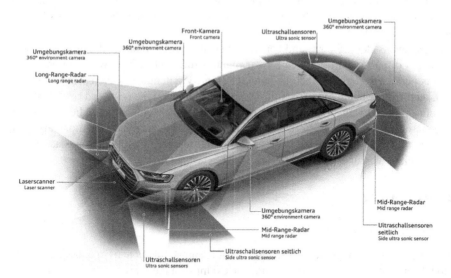

Fig. 1. Car sensor fields for environment monitoring [1].

In addition to the components of the AS, there are many more sensors in various areas of a vehicle that generate data through the operation of the car. This data is aggregated for further processing, e.g., to create value by data-driven business models.

For example, the business model of Usage-based Insurance is operated by many car insurance companies and is known as "Telematics Tariff". The amount of the premium to be paid is calculated on the basis of the kilometers travelled and the driving style of the policyholder [15]. The driving style is analyzed and evaluated by defined parameters and a scoring method [36]. Driver scoring is also used to sensitize the drivers of a fleet to prudent and cost-saving driving behavior [32]. Connected cars generate a lot of data about the driver [35]. This includes, for example, the use of the audio system or the login information of service access. The business model of Targeted Advertising aims at advertising revenues based on user preferences derived from this data. In the business model of Destination Prediction, the connected car uses data to suggest possible destinations to the driver [31].

Remote Diagnosis or Maintenance Management is a business model based on technical vehicle data. For example, the status of the engine is permanently sent to the car manufacturer or the corresponding service centers. As a result, the status of the vehicle is continuously evaluated and monitored. The business model of Vehicle Tracking uses the vehicle position data for monitoring and theft protection. E.g., by geofencing a signal is emitted when a vehicle leaves a defined area [3].

Environmental data can be used for the business model of traffic optimization. The combination of position data with speed data of a vehicle enables real-time traffic information. The sensors of the driver assistance systems continuously map the environment and generate information on road conditions, among other things. This information is used within the business model of Road Segment Data (RSD) that supplements online road maps for other road users. This idea of selling data of road conditions, generated by vehicle sensors, is the sample scenario in Sect. 4 to illustrate the corresponding risk assessment of data transfer.

In order to identify costs and benefits of such business models, the monetary value of data is of particular relevance for car manufacturers [19]. The two main objectives of this perspective are, on the one hand, to identify potentials of increasing efficiency and, on the other hand, to uncover possibilities of reducing the costs and risks [18]. The benefit-oriented evaluation of car data in particular can provide initial indications of the risks associated with data-driven business models. When calculating the monetary value of data, this includes expected values of potential losses, e.g., due to poor data quality or inefficient data use.

4 Risk Assessment

4.1 Risk Assessment Model

Figure 2 outlines a model for assessing risks of car data transfer and car data sharing. The model application is demonstrated using the example of the sale of road segment data. In this business model, vehicle-generated data is used to identify road damage and obstacles and to make it available to third parties (navigation map providers, suppliers, marketing agencies, insurance companies, etc.). In this use case data is generated by camera sensors (see Fig. 1) permanently recording during the journey of the vehicle. The aggregated data package includes edge markings, center markings, strip width, crash barriers, guide posts, signs, wild warning reflectors, and barriers, just to mention a few elements. The assessment process comprises three stages starting with "Identification" followed by "Analysis" and ending with "Evaluation" of the objects at risk (see Fig. 2). These three phases are illustrated by assessing the data transfer risks as one of various dimensions of the data-based business model.

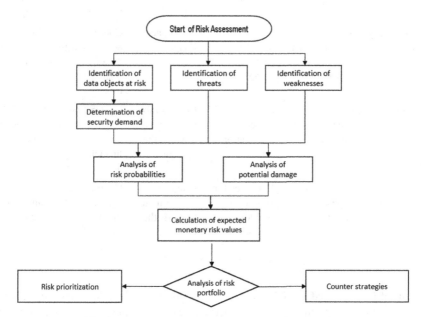

Fig. 2. Process of risk assessment for car data transfer.

4.2 Risk Assessment of Data Transfer

Identification of Data Transfer Risks. The vehicle-generated road segment data is the central data object at risk. Specifically, this is a combination of vehicle-generated and navigational position data. The RSD is evaluated in terms of its impact on principal risk types (see Table 1).

Table 1. Risk identification of RSD.

Risk type	Financial loss	Operating costs	Loss of reputation	Law & regulations	Health & safety
Confidentiality	Significant impact, as data is publicly available	Noticeable increase in customer complaints	No impact on image	Contractual penalties with minor financial losses	No injuries to persons
Integrity	Average losses due to unusable data records	Costs of inspection and elimination of errors	Considerable loss of image towards the customer	Contractual penalties with minor financial losses	No injuries to persons
Availability	Considerable losses due to missing data sets	Expenses for short term problem solution	Considerable loss of image towards the customer	Contractual penalties with minor financial losses	No injuries to persons

Analysis of Data Transfer Risks. Once threats have been identified, they are analyzed in terms of their probability of occurrence and extent of damage. To determine the probabilities, a Monte-Carlo simulation based frequency analysis is done (see Table 2). The method uses the OEM specified metric of Table 2 for each risk type in Table 1. It delivers a probability distribution showing the likelihood of each occurring risk (see Table 3).

Table 2. Metric of risk levels.

Probabilities of occurrence	Frequencies	Risk level
Most likely (99.99%)	Once a year or more often	a
Probably (30%)	Once in 1–5 years	b
Occasionally (10%)	Once in 5–25 years	c
Unlikely (5%)	Once in 25–100 years	d
Highly unlikely (1%)	Once in 100 years	e

Table 3. Probabilities of threats to RSD transfer.

Threat	Probabilities
Loss of information through partners	10%
Software error	30%
Manipulation of data records	5%
Other use of the data	10%
Unauthorized access to information by third parties	1%
Denial of service attack [40]	10%

The identified risks (see Table 1) and threats (see Table 3) lead to a combined risk assessment [24]. For a RSD business case costs of technical data losses, costs of reputation loss as well as economical sales losses are estimated (see Table 4).

Table 4. Analysis of monetary risks to RSD transfer.

Risk	Extent of damage
Loss of data records before anonymization	45.000.000 €
Loss of data records after anonymization	22.500.000 €
Incorrect data aggregation	850.000 €
Missed profits	27.500.000 €
Changing data records	4.000.000 €
Loss of sales volume	1.000.000 €
Paralyzing the cloud	1.000.000 €

Evaluation of Data Transfer Risks. In the final step of the risk assessment, the frequency and amount of potential damage resulting from risks are brought face to face.

This risk landscape in Fig. 3 is used for risk prioritization [4]: Disaster risk (A), survival risk (B), large risk (C), medium risk (D), small risk (E), very small risk (F). It reflects priority levels, with risks threatening the existence of the company being represented by level "A" and negligible risks by level "F".

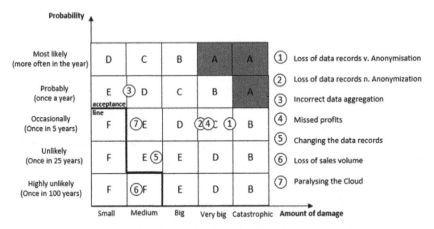

Fig. 3. Risk landscape of the RSD.

5 Conclusions

Connected cars generate a large amount of data that can be used for optimization or innovation of business models [3]. In this paper risks associated with these business models are addressed and focused on sensor generated data to be sold and transferred to third parties. The research follows three research questions:

1. *Which aspects could a risk assessment model for vehicle-generated data take into account?*

Risks can be characterized qualitatively as well as quantitatively [16]. In traditional approaches qualitative risk assessment is done primarily to identify risk categories and to describe them roughly [5, 12]. Methods of qualitative risk assessment can be transferred easily to intangible goods like car data. In the newly created model this takes place during the identification phase of the assessment process. However for quantitative risk analysis you have to take into account much more detailed technical aspects of data gathering, processing, and delivering. This is reflected in the analysis phase of the model. Here, different risk dimensions are quantitatively assessed. Among others statistical and scoring methods are applied to calculate probabilities, to classify damages and to estimate monetary losses. In the final evaluation phase the different risk dimensions are cross-validated. This results in a risk ranking. Based on the ranking the risks are prioritized for taking strategic counter-measures. The introduced aspects are not exhaustive but are significant risk elements of data-driven business models and selling data as a product is one of them.

2. *How does the application of this model look like in a practical example?*

The selected use case of assessing the risk of RSD sale gives partial insights into illustrative approaches to identify the crucial risk types of RSD, to analyze the probability and extent of damage related to improper technical data handling. Aggregating the risks in a multidimensional view the result shows that the loss of data records and suboptimal yield of data sale lead to very big damages which occur occasionally. Incorrect aggregation of RSD occurs most likely, but produces small to medium damage. The monetary damage resulting from the loss of data is the highest and outpaces the monetary losses of all other risks by the factor of two.

To sum up, it is shown that on the one hand vehicle data have a considerable potential for increasing the yield, but on the other hand they are fraught with risks. There are no approaches substantially described in the scientific literature that comprehensively propose risk management methods for data-driven business models in the automotive industry. The core of a newly customized model is based on the fact that business models are to be evaluated from four perspectives: data protection and security, economic efficiency, customer satisfaction, and personal security. In this paper, the primary view is on data protection and data security as well as profitability. In further work, therefore, the areas of customer satisfaction and the security of persons have to be investigated and examined in the context of the risk assessment of data-based business models.

References

1. AUDI AG: Sensorfelder der Umfeldüberwachung. https://www.audi-mediacenter.com/de/tec hnik-lexikon-7180/fahrerassistenzsysteme-7184
2. Baheti, R., Gill, H.: Cyber-physical systems. Impact Control Technol. **12**(1), 161–166 (2011)
3. Bosler, M., Burr, W., Ihring, L.: Vernetzte Fahrzeuge–empirische Analyse digitaler Geschäftsmodelle für Connected-Car-Services. HMD Praxis der Wirtschaftsinformatik **55**(2), 329–348 (2018). https://doi.org/10.1365/s40702-018-0396-8
4. Brünger C.: Nutzenkonsistente Risikopriorisierung: Die Risk-Map im Kontext rationaler Entscheidungen, 1st edn. Paderborn, Univ., Diss., Zugl. (2011). Gabler Verlag/Springer Fachmedien Wiesbaden GmbH Wiesbaden, Wiesbaden (2011)
5. Chapman, C., Ward, S.: Project Risk Management: Processes, Techniques and Insights, 2nd edn. Wiley, Hoboken (2003)
6. Diederichs, M.: Risikomanagement und Risikocontrolling. Vahlen (2017)
7. Dijkman, R.M., Sprenkels, B., Peeters, T., Janssen, A.: Business models for the Internet of Things. Int. J. Inf. Manage. **35**(6), 672–678 (2015)
8. ElBatt, T., Saraydar, C., Ames, M., Talty, T.: Potential for intra-vehicle wireless automotive sensor networks. In: 2006 IEEE Sarnoff Symposium, pp. 1–4. IEEE, Newark (2006)
9. Fleisch, E., Weinberger, M., Wortmann, F.: Business models and the Internet of Things (extended abstract). In: Podnar Žarko, I., Pripužić, K., Serrano, M. (eds.) Interoperability and Open-Source Solutions for the Internet of Things. LNCS, vol. 9001, pp. 6–10. Springer, Cham (2015). https://doi.org/10.1007/978-3-319-16546-2_2
10. Fleming, W.J.: New automotive sensors—a review. IEEE Sens. J. **8**(11), 1900–1921 (2008)

11. Fleming, W.J.: Overview of automotive sensors. IEEE Sens. J. **1**(4), 296–308 (2001)
12. Gladen, W.: Performance Measurement: Controlling mit Kennzahlen, 6th edn. Springer Gabler, Wiesbaden (2014). https://doi.org/10.1007/978-3-658-05138-9
13. Hesse, J., Gardner, J.W., Göpel, W.: Sensors for Automotive Technology, 1st edn. Wiley-VCH Verlag, Weinheim (2003)
14. Hornung, G.: Verfügungsrechte an fahrzeugbezogenen Daten. Datenschutz und Datensicherheit - DuD **39**(6), 359–366 (2015). https://doi.org/10.1007/s11623-015-0430-8
15. Johanning, V., Mildner, R.: Car IT kompakt. Springer, Wiesbaden (2015). https://doi.org/10.1007/978-3-658-09968-8
16. Jorion, P.: Value at Risk, 3rd edn. McGraw-Hill Education Ltd., New York City (2006)
17. Ju, J., Kim, M.S., Ahn, J.H.: Prototyping business models for IoT service. Procedia Comput. Sci. **91**, 882–890 (2016)
18. Kleindorfer, P.R., Saad, G.H.: Managing disruption risks in supply chains. Prod. Oper. Manage. **14**(1), 53–68 (2005)
19. Krotova, A., Rusche, C., Spiekermann, M.: Die ökonomische Bewertung von Daten: Verfahren, Beispiele und Anwendungen. No. 129, IW-Analysen (2019)
20. Lee, E.A.: Cyber-physical systems-are computing foundations adequate. In: Position PAPER for NSF Workshop on Cyber-Physical Systems: Research Motivation, Techniques and Roadmap, vol. 2, pp. 1–9. Citeseer (2006)
21. Leminen, S., Westerlund, M., Rajahonka, M., Siuruainen, R.: Towards IOT ecosystems and business models. In: Andreev, S., Balandin, S., Koucheryavy, Y. (eds.) NEW2AN/ruSMART - 2012. LNCS, vol. 7469, pp. 15–26. Springer, Heidelberg (2012). https://doi.org/10.1007/978-3-642-32686-8_2
22. Loebbecke, C., Picot, A.: Reflections on societal and business model transformation arising from digitization and big data analytics: a research agenda. J. Strateg. Inf. Syst. **24**(3), 149–157 (2015)
23. Madden, S., Franklin, M.J., Hellerstein, J.M., Hong, W.: TAG: a tiny aggregation service for ad-hoc sensor networks. ACM SIGOPS Oper. Syst. Rev. **36**(SI), 131–146 (2002)
24. McNeil, A.J., Frey, R., Embrechts, P.: Quantitative Risk Management: Concepts. Techniques and Tools. Princeton University Press, Princeton (2015). Revised ed.
25. Mikusz, M., Jud, C., Schäfer, T.: Business model patterns for the connected car and the example of data orchestrator. In: Fernandes, J.M., Machado, R.J., Wnuk, K. (eds.) ICSOB 2015. LNBIP, vol. 210, pp. 167–173. Springer, Cham (2015). https://doi.org/10.1007/978-3-319-19593-3_14
26. Morris, D., Madzudzo, G., Garcia-Perez, A.: Cybersecurity and the auto industry: the growing challenges presented by connected cars. Int. J. Automot. Technol. Manage. **18**(2), 105–118 (2018)
27. Muhtaroğlu, F.C.P., Demir, S., Obalı, M., Girgin, C.: Business model canvas perspective on big data applications. In: 2013 IEEE International Conference on Big Data, Santa Clara, CA, USA, pp. 32–37. IEEE (2013)
28. Nasr, E., Kfoury, E., Khoury, D.: An IoT approach to vehicle accident detection, reporting, and navigation. In: 2016 IEEE International Multidisciplinary Conference on Engineering Technology (IMCET), Beirut, Lebanon, pp. 231–236. IEEE (2016)
29. Naumov, V., Gross, T.R.: Connectivity-aware routing (CAR) in vehicular ad-hoc networks. In: IEEE INFOCOM 2007-26th IEEE International Conference on Computer Communications, Glasgow, Scotland, pp. 1919–1927. IEEE (2007)
30. Porter, M., Heppelmann, J.: How smart, connected products are transforming companies. Havard Bus. Rev. **93**(10), 96–112 (2015)
31. Proff, H., Fojcik, T.M.: Mobilität und digitale Transformation - Technische und betriebswirtschaftliche Aspekte - Schlussbetrachtung. Mobilität und digitale Transformation, pp. 609–611. Springer, Wiesbaden (2018). https://doi.org/10.1007/978-3-658-20779-3_38

32. Prytz, R.: Machine Learning Methods for Vehicle Predictive Maintenance Using Off-Board and On-Board Data, 1st edn. Halmstad University Press, Halmstad (2014)
33. Russell, M.E., Drubin, C.A., Marinilli, A.S., Woodington, W.G., Del Checcolo, M.J.: Integrated automotive sensors. IEEE Trans. Microw. Theory Tech. **50**(3), 674–677 (2002)
34. Shah, S.H., Yaqoob, I.: A survey: Internet of Things (IOT) technologies, applications and challenges. In: 2016 IEEE Smart Energy Grid Engineering (SEGE), Oshawa, Ontario, Canada, pp. 381–385. IEEE (2016)
35. Shahraray, B., et al.: U.S. Patent No. 9,403,482. U.S. Patent and Trademark Office, Washington, DC (2016)
36. Streich, M., D'imperio, A., Anke, J.: Bewertung von Anreizen zum Teilen von Daten für digitale Geschäftsmodelle am Beispiel von Usage-based Insurance. HMD Praxis der Wirtschaftsinformatik **55**(5), 1086–1109 (2018)
37. Stulz, R.M.: Rethinking risk management. J. Appl. Corp. Financ. **9**(3), 8–25 (1996)
38. Westbrook, M.H.: Automotive Sensors, 1st edn. Momentum Press, New York (2009)
39. Westerlund, M., Leminen, S., Rajahonka, M.: Designing business models for the internet of things. Technol. Innov. Manage. Rev. **4**, 5–14 (2014)
40. Wood, A.D., Stankovic, J.A.: Denial of service in sensor networks. Computer **35**(10), 54–62 (2002)

Review for Message-Oriented Middleware

Yang Liu[1,2,3,4(✉)], Liang-Jie Zhang[3,4], and Chunxiao Xing[1,2]

[1] Research Institute of Information Technology, Beijing National Research Center for Information Science and Technology, Tsinghua University, Beijing 100084, China
1491701161@qq.com
[2] Department of Computer Science and Technology, Institute of Internet Industry, Tsinghua University, Beijing 100084, China
[3] National Engineering Research Center for Supporting Software of Enterprise Internet Services, Shenzhen 518057, China
[4] Kingdee Research, Kingdee International Software Group Company Limited, Shenzhen 518057, China

Abstract. With the rapid development of 5G technology and micro-service software developing technology, a growing number of intelligent devices are connected to the Internet, and it has become a trend to provide services to customers in a collaborative way. As the information exchange center of collaborative services, the role of message-oriented middleware becomes more and more important. Firstly, this paper describes the definition, main characteristics, core technology, main products and existing problems of middleware. Then, it takes one of the problems as an example, designs a gateway architecture based on a message middleware, and introduces its application in the Internet of things in detail.

Keywords: Middleware (MOM) · Message queue · Publish/Subscribe · IoT

1 Overview of MOM

1.1 Definition

The traditional communication mode is remote procedure call mode (RPC). In this mode (see Fig. 1), applications are highly dependent and synchronous [1]. What are the disadvantages of such a model? For example, in transaction, the order system will send messages to the logistics system and the notification system. If notification system is broken, messages to be sent to it will pile up all the time in the order system and the whole system will crash. Message-oriented middleware (MOM) can solve this problem, which is a kind of technology that uses message delivery mechanism or message queue mode for data exchange and multi system integration [2, 3].

1.2 Main Feature and Applying Patterns

The main features of MOM include the following seven aspects, asynchronous transmission, defense communication, concurrent execution, log communication,

© Springer Nature Switzerland AG 2020
W. Song et al. (Eds.): ICIOT 2020, LNCS 12405, pp. 152–159, 2020.
https://doi.org/10.1007/978-3-030-59615-6_12

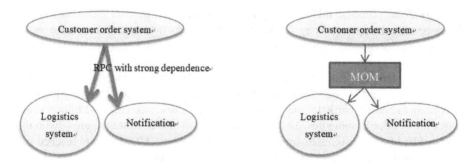

Fig. 1. RPC vs MOM.

multiple communication modes, application isolation from network complexity, and flow elasticity control [4]. These typical characteristics can be seen from its two applying modes (see Fig. 2). Mode 1 can reflect the first six characteristics of middleware technology, and mode 2 can help us understand the seventh, flow elastic control. In the first two times of message flow, the time delay is relatively large, 30 ms and 300 ms. Later, middleware technology is used, and the time delay is reduced to 1 ms, which is because it is used to control network traffic.

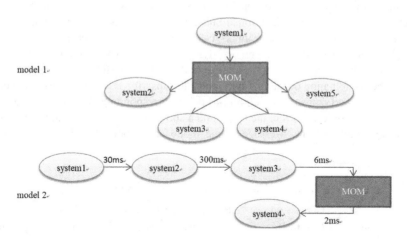

Fig. 2. Applying models of MOM.

1.3 Developments in MOM

The concept of middleware technology has a long history, but it has been widely used in the last decade. The first middleware technology is CICS developed by IBM [5], and the first distributed middleware is tuxedo [6], developed by Bell Labs in 1984. Subsequently, Some MOM products, Microsoft' MSMQ, Apache' ActiveMQ, appear in 2003 [7]. In 2007, RabbitMQ 1.0 [8], developed by rabbit technologies based on AMQP standard, was released. In 2010, LinkedIn developed its own messaging system

Kafka [9]. The domestic middleware started later. The first product was TongLNK/Q developed by Dongfang Tong Technology Co., Ltd. in 1995. Then, some other excellent products came out successively, such as BusinessMQ in 2016, RocketMQ in 2017 [10], etc.

With the development of Internet of things, cloud computing, block-chain and other technologies [11, 12], the global MOM market has grown rapidly in recent years. According to Gartner, in 2018, the global middleware market exceeded 32 billion yuan, with a growth rate of more than 12.5%. China's market also continues to grow, with the needs of building new infrastructure. In 2018, the domestic MOM market reached 6.5 billion, and in 2019, it reached 7.24 billion, with a growth of more than 10%. In the next few years, MOM will continue to maintain a steady growth momentum.

2   Core Technology

2.1   Communication Mode in MOM

Message middleware can store and forward messages. How does it work internally? It has three working modes: point-to-point mode, publish/subscribe mode and message queue mode [13].

Differences between the first two are obvious (see Table 1). (1) Point to Point is to maintain a logical link for two applications. After a message is consumed, it is no longer stored in the queue. (2) Publish/Subscribe is that messages are published to a specified topic, and multiple consumers can subscribe to the same messages in that topic [14]. This mode has advantages of dynamic routing, asynchronous transmission and fault tolerance, and thus become an informal standard of MOM. (3) Message queue mode is the combination of the first two. Publish/Subscribe is adopted between queues, but the queues and consumers are point-to-point corresponding. That is, messages are broadcast to multiple queues in the topic, and a subscription group consumes messages in a queue point-to-point. Message queue mode can support multiple requirements more flexibly.

Table 1. Comparison of communication modes.

Mode	Correspondence	Pattern of consumption
Point to Point	**One to One**	Consumers take the initiative to pull messages, and only one consumer can get a message
Publish/Subscribe	**One to Many**	One message is pushed to all consumers

2.2   MOM Based on Message Queue

Architecture of MOM. The MOM based on message queue mode consists of four parts (see Fig. 3), interface processing module, queue manager, message channel agent and security management [15]. Among them, the interface processing module is used to process service requests from various applications, to separate or combine data streams according to the types of requests. Queue manager is the core of MOM, which is responsible for creating and deleting queues, controlling its mode and priority. The message channel agent uses the network to provide the communication mechanism, which is responsible for delivering messages to the responsible queue, monitoring the channel failure and dealing with the failure in time. Security management mainly completes the encryption/decryption of data and provides users with secure messages.

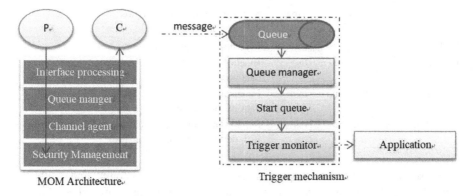

Fig. 3. MOM architecture and trigger mechanism.

Trigger Mechanism of Message. When the application accessing middleware is in continuous running state, any arriving message can be read immediately, but this is only applicable when the message flow is stable. In most cases, the application is not active before the message arrives. In this case, the trigger mechanism of message queue needs to be set to activate the dormant application. Queue manager and Channel agent are combined to perform message triggering (see Fig. 3). (1) A message arrives at the message queue. (2) The queue manager references the environment information to determine whether the arrival of this message constitutes a trigger event. (3) The queue manager creates a trigger message and sends it to the start queue. (4) The trigger monitor reads the trigger message from the start queue and issues an activation command. (5) After the application is activated, the message is taken from the trigger queue and the corresponding operation is performed.

In order to achieve a higher level of response speed, MOM often needs to combine Redis or other caching technologies to meet specific business needs, such as the commodity Seckill system. The following is the applicable scenario analysis of various MOMs.

3   Popular MOM and Performance Comparison

At present, popular MOM products are mainly Kafka, RabbitMQ, RocketMQ, ActiveMQ. Their throughput, delay of forwarding messages and suitable application scenarios are shown in the table. Users can select appropriate middleware according to their own scenario requirements (see Table 2).

Table 2.   Product features comparison.

Product	Throughput	Time delay	Suitable application scenarios
Kafka	100 thousand	ms	Big capacity data acquisition
RabbitMQ	10 thousand	us	Business interaction with high requirements for data consistency, stability and reliability
RocketMQ	100 thousand	ms	Financial field with large amount of transaction data in short term
ActiveMQ	10 thousand	ms	Support multiple protocol access and high real-time IOT device interaction scenarios

These MOMs basically implement models of "point-to-point" and "publish/subscribe" mentioned above, and their important differences are mainly reflected in the way consumers consume messages. Kafka is a pull model, and RabbitMQ is a push model, while RocketMQ and ActiveMQ both implement the combination of push and pull. These two consumption models have their own advantages. The push model highlights the real-time nature of messages, while the pull model is convenient for the control of consumption rhythm.

4   Existing Issues

Message middleware has become a key part of modern data-driven architecture, but under the requirements of new environment and complex scenarios, there are still some problems to be solved in the field of message communication.

(1) There is not a set of industry standards that have nothing to do with suppliers and development languages. The internal implementation of various message middleware is not uniform and compatible with each other. In order to ensure the normal communication between message engines, users have to invest a lot of energy to deal with compatibility issues, which will cause higher access costs and maintenance costs. Therefore, the establishment of factual standards in the field of information communication has become the common appeal of developers and relevant partners.

(2) A large number of messages are piled up in the queue, resulting in serious delay of the client. In the industrial Internet, the real-time requirement of device-level messages is very high [14–16]. When one device fails and can't accept messages, the queue will pile up a large number of messages, and other devices can't get response in time.

5   Application of MOM in IoT

As mentioned in Sect. 4, an existing issue is that if MOM is piled up with large number of messages, serious time delay may occurs in IoT device. This section designed a IoT message system to this problem.

5.1   Architecture of IoT Message System

The Internet of things (IoT) message system is composed of IoT device nodes and message gateways. Distributed deployment is adopted in the local area network, and data communication is carried out between nodes through messages. The system architecture is shown in Fig. 4. After the IoT node collects the environment data, it encapsulates the data into message format and sends the message to the gateway through the client program. The core of the gateway is the ActiveMQ server. With it, the gateway can accept data from nodes. After cleaning, analyzing and filtering the data format, it transmits the data to the cloud server through MQTT protocol. In theory, a device node can publish messages to multiple topics of the middleware server, and multiple nodes can also publish messages to the same topic. In order to avoid the problem of mutual influence between devices caused by message accumulation in gateways, this system adopts the design of device-level topics.

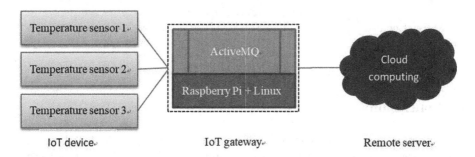

Fig. 4.   Architecture of IoT system.

5.2   Main System Functions

Device Connection. The IoT device consists of mini PC Raspberry Pi and intelligent temperature sensor DS18B20. Using the metal pins on the development board, the sensor can be plugged into the mini PC. After the hardware is connected, run the Linux operating system on the PC, and then install the software stomp.py. This software provides functions based on simple text protocol, which can encapsulate the data collected by the sensor into formatted messages. The hardware used for the gateway can also be the Raspberry Pi, which also runs the Linux, on which the message server

ActiveMQ is installed. Sensor nodes establish network connection with gateway through WiFi, and gateway connects with remote cloud computing platform through Internet service.

Function Design. This architecture uses ActiveMQ to design device-level message queue to store and forward device related messages (see Fig. 5). Several topics are set in the gateway server. One topic receives the temperature data of a sensor node. The data acquisition program on the device is developed in Python to obtain the temperature data of the environment. Then methods of the STOMP library are called to perform format conversion, encapsulate the data into message format and send it to the message queue of the gateway. As the client of ActiveMQ server, the remote cloud computing service center obtains data from multiple gateways. After data aggregation, the cloud center obtains some decisions by analyzing these data. According to the decision, the center sends backs messages to manage the devices, and in turn adjusts the operation status of those IoT devices.

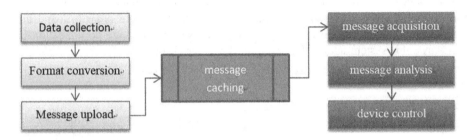

Fig. 5. Functional flows.

6   Conclusion

This paper introduces technology of MOM and its application in the IoT. Users can have a comprehensive understanding of it according to these introductions, and select appropriate middleware products to build their own message system. The future research direction mainly includes three aspects. (1) How to combine the new distributed communication standard with MOM? In distributed service architecture, there are many application-level protocols. MOM should support the function of multi-protocol communication, and need to improve this function module constantly. (2) How to be compatible with multiple middleware solutions of different communication modes? The compatibility of multiple middleware can make them give full play to their maximum performance in 5G big data, IoT and other new environments. (3) How to combine middleware with block-chain? In the collaborative process of multi system, a lot of data needs to be saved in several systems. This is in line with the idea of decentralization in block-chain. So, applying block-chain to secure middleware system is bound to become a trend.

References

1. Kumar, N., Kumar, A., Giri, S.: Design and implementation of Three Phase Commit Protocol directory structure through Remote Procedure Call application. In: International Conference on Information Communication and Embedded Systems, Chennai, pp. 1–5. IEEE (2014)
2. Li, Q., Zhou, M.: The state of the art in middleware. In: International Forum on Information Technology and Applications, Kunming, China, pp. 83–85. IEEE (2010)
3. Bae, Y., Oh, B., Moon, K., Ha, Y., Kim, S.: Design and implementation of an adaptive middleware based on the universal middleware bridge for heterogeneous home networks. IEEE Trans. Consum. Electron. **56**(2), 620–626 (2010)
4. Razzaque, M.A., Milojevic-Jevric, M., Palade, A., Clarke, S.: Middleware for internet of things: a survey. IEEE Internet Things J. **3**(1), 70–95 (2016)
5. Bainbridge, A., Colgrave, J., Colyer, A., Normington, G.: CICS and enterprise JavaBeans. IBM Syst. J. **40**(1), 46–67 (2001)
6. Altiok, T., Xiong, W., Gunduc, M.: A capacity planning tool for the Tuxedo middleware used in transaction processing systems. In: Proceeding of the 2001 Winter Simulation Conference (Cat. No.01CH37304), Arlington, VA, USA, pp. 502–507. IEEE (2001)
7. Chun, J.-K., Cho, S.-H.: Performance and stability testing of MSMQ in the .NET environment. In: Third International Workshop on Electronic Design, Test and Applications (DELTA), Kuala Lumpur, pp. 496–502. IEEE (2006)
8. Ionescu, V.M.: The analysis of the performance of RabbitMQ and ActiveMQ. In: 14th RoEduNet International Conference - Networking in Education and Research (RoEduNet NER), Craiova pp. 132–137. IEEE (2015)
9. Chai, X.-C., Wang, Q.-L., Chen, W.-S., Wang, W.-Q., Wang, D.-N., Li, Y.: Research on a distributed processing model based on Kafka for large-scale seismic waveform data. IEEE Access **8**, 39971–39981 (2020)
10. Yue, M., Ruiyang, Y., Jianwei, S., Kaifeng, Y.: A MQTT protocol message push server based on RocketMQ. In: 10th International Conference on Intelligent Computation Technology and Automation (ICICTA), Changsha, pp. 295–298. IEEE (2017)
11. Hiraman, B.R., Chapte Viresh, M., Karve Abhijeet, C.: A study of Apache Kafka in big data stream processing. In: International Conference on Information, Communication, Engineering and Technology (ICICET), Pune, pp. 1–3. IEEE (2018)
12. da Cruz, M.A.A., Rodrigues, J.J.P.C., Al-Muhtadi, J., Korotaev, V.V., de Albuquerque, V. H.C.: A reference model for internet of things middleware. IEEE Internet Things J. **5**(2), 871–883 (2018)
13. Sheltami, T., Al-Roubaiey, A., Mahmoud, A., Shakshuki, E.: A publish/subscribe middleware cost in wireless sensor networks: a review and case study. In: 28th Canadian Conference on Electrical and Computer Engineering (CCECE), Halifax, NS, pp. 1356–1363. IEEE (2015)
14. Al-Roubaiey, A.A., Sheltami, T.R., Hasan Mahmoud, A.S., Salah, K.: Reliable middleware for wireless sensor-actuator networks. IEEE Access **7**, 14099–14111 (2019)
15. Yu, D., Park, H.S.: Real-time middleware with periodic service for industrial robot. In: 14th International Conference on Ubiquitous Robots and Ambient Intelligence (URAI), Jeju, South Korea, pp. 879–881. IEEE (2017)
16. Justin Dhas, Y., Jeyanthi, P.: A review on internet of things protocol and service oriented middleware. In: International Conference on Communication and Signal Processing (ICCSP), 2019, Chennai, India, pp. 0104–0108. IEEE (2019)

Author Index

Aleman, Febe Hernandez   125
Allan, Moray   45
Ayoub, Raid   63

Bodendorf, Frank   141

Cherkasova, Ludmila   45, 63

Ergun, Kazim   63

Franke, Jörg   141
Fry, John   45

Gandomi, A. H.   96

Huang, Yan   1

Jeyakumar, Jeya Vikranth   45
Jung, Markus   82

Krishnamoorthy, Rahul   133

Lajevardi, Saina   45
Le, Franck   32
Lee, Dongha   82
Li, Wei   1
Liu, Rex   133
Liu, Xin   133
Liu, Yang   152
Lopez-de la Fuente, Martha S.   125

Mercati, Pietro   63
Mishra, Deepak   108

Nagesh, Nitish   63

Omidvar, Sorush   108

Pang, Junjie   1
Park, Jiye   82

Ramasubbareddy, Somula   96
Ramaswamy, Lakshmish   108
Ramli, Albara Ah   133
Rathgeb, Erwin P.   82
Raza, Usman   17
Rosing, Tajana   63

Salam, Abdul   17
Samanta, Sidharth   96
Sankar, S.   96
Setayeshfar, Omid   108
Singhar, Sunil Samanta   96
Srivastava, Mani   45
Srivatsa, Mudhakar   32

Tagkopoulos, Ilias   133
Tonekaboni, Navid Hashemi   108

Vishal, I. B.   133

Wang, Xiaoxiao   133

Xie, Zhenzhen   1
Xing, Chunxiao   152

Yang, Jishen   1
Yu, Xiaofan   63

Zhang, Liang-Jie   152
Zhao, Yue   45

Printed in the United States
By Bookmasters